Applied ethology 2017

Understanding animal behaviour

ISAE2017

Proceedings of the 51st Congress of the International Society for Applied Ethology

7-10 August 2017, Aarhus, Denmark

Understanding animal behaviour

edited by:

Margit Bak Jensen

Mette S. Herskin

Jens Malmkvist

Online Academic Submission and Evaluation System

The sculpture of the pigs on the front cover is called 'Grisebrønden' (meaning 'The pig well' in English) which is placed at the town hall square in Aarhus

EAN: 9789086863112
e-EAN: 9789086868582
ISBN: 978-90-8686-311-2
e-ISBN: 978-90-8686-858-2
DOI: 10.3920/978-90-8686-858-2

Photo cover:
Maria V. Rørvang, Aarhus University

First published, 2017

© Wageningen Academic Publishers
The Netherlands, 2017

Wageningen Academic
P u b l i s h e r s

Welcome to the 51st Congress of ISAE

Understanding animal behaviour is the overall theme of this congress and the red thread through the chosen scientific topics. This reflects our wish for focus on what we consider the core of applied ethology. Understanding animal behaviour is essential in order to improve the interaction between animals and the environments in which they are kept and to improve animal welfare. We need to understand how the animals sense and perceive their world beyond the human perspective, which is based almost only on vision. Animal affective states are central for the concept of animal welfare, but to advance our assessment of these states we have to understand animals' learning and cognitive abilities. The potential, but also pitfalls, of novel methods to study behavioural and physiological stress responses must be addressed continuously. Many of the animals kept by man are social, and we need to understand social behaviour in relation to their welfare, and especially because we rely on social behaviour when animals are group housed. In addition, for many domestic animal species we rely on maternal behaviour for neonatal development and survival. Animal behaviour is a key component in ensuring animal welfare and functionality in extensively and intensively kept animals and 'applying ethology in the keeping of animals', thus putting the science to work is by number the largest topic on this congress. We warmly welcome you to Aarhus and hope that you will enjoy the scientific program and discussions as well as the social events and networking. Please, help us to make this congress a memorable event.

Margit Bak Jensen, Jens Malmkvist and Mette S. Herskin

Acknowledgements

Congress Organising Committee
Margit Bak Jensen (chair), Mette S. Herskin, Lene Juul Pedersen, Lene Munksgaard, Jens Malmkvist, Karen Thodberg, Claus Bo Andreasen, Tina Albertsen (all Aarhus University)

Scientific Committee
Margit Bak Jensen (chair), Mette S. Herskin, Lene Juul Pedersen, Lene Munksgaard, Jens Malmkvist, Karen Thodberg (all Aarhus University) Jan Ladewig, Björn Forkman (both University of Copenhagen)

Social Committee
Karen Thodberg, Lene Juul Pedersen

Media
Claus Bo Andreasen, Nina Hermansen (both Aarhus University)

Ethics Committee
Anna Olsson (Chair), Portugal; Francesco De Giorgio, The Netherlands; Francois Martin, USA; Cecilie Mejdell, Norway; Franck Peron, France; Elize van Vollenhoven, South Africa; Alexandra Whittaker, Australia

Technical support
Jens Bech Andersen, Lars Bilde Gildberg, Mads Ravn Jensen (all Aarhus University)

Secretary assistance
Tina Albertsen

ISAE helpers
Katrine Kop Fogsgaard, Sarah-Lina Aagaard Schild, Mona Lilian Vestbjerg Larsen, Maria Vilain Rørvang, Fernanda Machado Tahamtani, Julie Cherono Schmidt Henriksen, Toke Munk Schou, Cecilie Kobek Thorsen, Claire Toinon, Anne S. Bak and Birthe Houbak (all Aarhus University)

Professional conference organisers
DIS Congress Service Copenhagen A/S (www.discongress.com)

Reviewers

Michael Appleby
Gareth Arnott
Harry Blokhuis
Xavier Boivin
Laura Boyle
Bjarne Braastad
Donald Broom
Stephanie Buijs
Oliver Burman
Andy Butterworth
Irene Camerlink
Janne W. Christensen
Sylvie Cloutier
Michael Cockram
Melanie Connor
Jonathan Cooper
Rick D'Eath
Antoni Dalmau
Ingrid De Jong
Trevor DeVries
Pierpaolo Di Giminiani
Laura Dixon
Rebecca Doyle
Cathy Dwyer
Bernadette Earley
Sandra Edwards

Inma Estevez
Emma Fabrega
Mark Farnworth
Katrine Kop Fogsgaard
Francisco Galindo
Lorenz Gygax
Derek Haley
Laura Hänninen
Moira Harris
Marie Haskell
Suzanne Held
Per Jensen
Jenna Kiddie
Ute Knierim
Paul Koene
Fritha Langford
Alistair Lawrence
Caroline Lee
Pol Llonch
Maja Makagon
Xavier Manteca
Jeremy Marchant-Forde
Georgia Mason
Rebecca Meagher
Suzanne Millman
David Morton

Ruth Newberry
Christine Nicol
Birte L. Nielsen
Per P. Nielsen
Cheryl O'Connor
Elizabeth Paul
Jean-Loup Rault
Bas Rodenburg
Jeffrey Rushen
Kenny Rutherford
Lars Schrader
Karin Schütz
Janice Siegford
Mhairi Sutherland
Fernanda Tahamtani
Michael Toscano
Cassandra Tucker
Simon Turner
Anna Valros
Heleen Van De Weerd
Antonio Velarde
Daniel Weary
Francoise Wemelsfelder
Christoph Winckler
Hanno Würbel

Sponsors

Main sponsor:

Bronze sponsors:

Exhibitor:

Maps

Welcome reception

Aarhus University's Main Hall and surrounds.
The address of the Main Hall (building 1412) is Nordre Ringgade 4, 8000 Aarhus C.

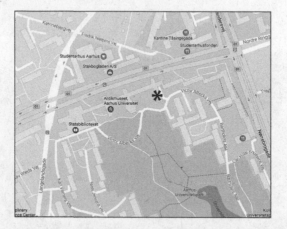

Main conference venue

Aarhus Music Hall and surrounds.
The address of the Music Hall is Thomas Jensens Allé 2, 8000 Aarhus C

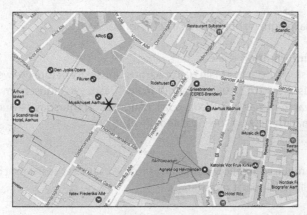

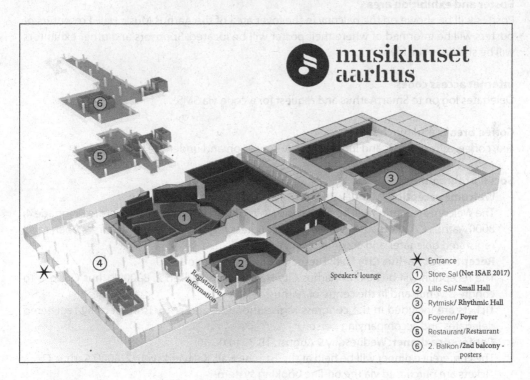

Speakers' lounge

* Entrance
1. Store Sal (Not ISAE 2017)
2. Lille Sal/ Small Hall
3. Rytmisk/ Rhythmic Hall
4. Foyeren/ Foyer
5. Restaurant/Restaurant
6. 2. Balkon /2nd balcony - posters

Registration/information

General information

Venues
Registration and welcome reception will be at the Main Hall, Aarhus University, Nordre Ringgade 4, 8000 Aarhus C.
The congress will be held at the Aarhus Music Hall, Thomas Jensens Allé 2, 8000 Aarhus C.

Official language
English is the official language of the ISAE2017 Congress.

Registration and information desk
The registration desk will be at the Main Hall at Aarhus University before the welcome reception on Monday 7 August (15.00 to 18.00) and additionally Tuesday 8 August from 8.00 at the Music Hall in the Foyer near the Small Hall. Wednesday 9 August and Thursday 10 August the opening hours of the information desk will be indicated on-site.

Name badges
Name badges are required to allow admittance to the congress sessions, coffee breaks and lunches. Badges will be issued at registration.

Poster and exhibition area
Posters will be shown on the balcony in the foyer area of the Aarhus Music Hall. Presenters of posters will be informed of where their poster will be located. Sponsors and other exhibitors will be shown in the foyer.

Internet access codes
Delegates log on to SmartAarhus and request for a code via SMS.

Coffee breaks and lunches
Tea, coffee, refreshments and lunches will be served on and under the balcony.

Social programme
- **Welcome reception**: Monday 7 August, 17.00 to 19.00
  The Welcome reception will be held at the Main Hall, Aarhus University, Nordre Ringgade 4, 8000 Aarhus C. Tickets are included in the congress registration fee, and entry is open to all registered delegates and accompanying persons.
- **Reception at Aarhus City Hall**: Tuesday 8 August, 17.30 to 18.30
  The reception will be held at Aarhus City Hall, Rådhuspladsen 2, 8000 Aarhus C (close to congress venue and in the centre of Aarhus).
  Tickets are included in the congress registration fee, and entry is open to all registered delegates and accompanying persons.
- **Conference dinner**: Wednesday 9 August, 18.30 to 00.30
  The conference dinner will be held at the Turbine Hall, Kalkværksvej 12, 8000 Aarhuc C. Tickets are purchased via the on-line booking system.
- **Farewell reception**: Thursday 10 August, 18.00 to 19:00
  The farewell reception will be held at Aarhus Music Hall, Thomas Jensens Allé 2, 8000 Aarhus C. Tickets are included in the congress registration fee, and entry is open to all registered delegates and accompanying persons.

How to get there

Aarhus University, Main Hall, Nordre Ringgade 4, 8000 Aarhus C:
The distance from Aarhus city centre (Aarhus train station) to the venue of the welcome reception is around 2.8 km. It takes around half an hour to walk to the venue. The venue is easily accessed by bus. Several buses run from in front of the Aarhus City Hall located close to the main train station. Please visit the ISAE homepage (www.ISAE2017.com) for more information.

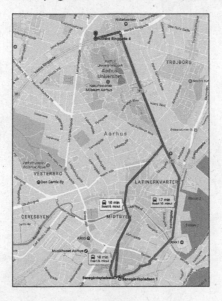

Aarhus Music Hall, Thomas Jensens Allé 2, 8000 Aarhus C:
The distance from Aarhus city centre (Aarhus train station) to the venue of the conference is around 500 m, and it will only take around 10 minutes to walk the distance.

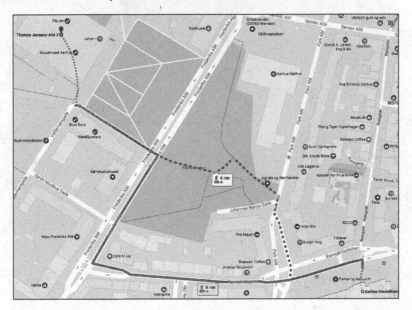

Scientific and social programme

Monday 7 August

15.00-18.00	Registration at Aarhus University, Main Hall

17.00-19.00	Welcome reception at Aarhus University. Venue: Main Hall

Tuesday 8 August

8.00	Placing of posters and registration ....

9.00	Opening Ceremony at Aarhus Music Hall. Venue: Rhythmic Hall

9.45	Wood-Gush Memorial Lecture: Peter Teglberg Madsen: Echolocating whales as models for studying how information flow governs behaviour. Venue: Aarhus Music Hall, Rhythmic Hall

10.30	Break ....

	Session 1, Topic 01 at Rhythmic Hall: The world of animals: senses and perception	Session 2, Topic 02 at Small Hall: Free papers
11.30	The role of olfaction in relation to stress and fear. *Vincent Bombail*	Literature review on priority welfare issues of commercially raised bison in North America. *John S. Church*
12.00	Olfactory enrichment for cattle? Development of tests for determining olfactory investigation of complex odours in cattle. *Maria Vilain Rørvang*	Associations between measures of rising and lying down behaviour in dairy cows. *Nina Dam Otten*
12.15	Do dogs see the Ponzo illusion? *Sarah-Elizabeth Byosiere*	Feedlot cattle are motivated to obtain roughage and show contrafreeloading. *Cassandra Tucker*
12.30	Lunch	

Tuesday 8 August

	Session 3, Topic 01 at Rhythmic Hall: The world of animals: senses and perception	Session 4, Topic 03 at Small Hall: Human-animal interactions
13.30	Taking into account different sensory abilities of animals in their housing and management. *Birte L. Nielsen*	Dogs, humans and genes: evidence for mutualistic symbiosis? *Per Jensen*
14.00	The impact of ultraviolet wavelength on broiler chicken welfare indicators. *Charlotte James*	Rat tickling: a systematic review of methods, outcomes and moderators. *Megan R. Lafollette*
14.15	How incline angel affects the preparatory step prior to climbing a ramp in domestic fowl? *Alexandra Harlander*	Cattle handling by untrained stockpeople gets worse along a vaccination work day. *Maria Camilla Ceballos*
14.30	Do mink mothers recognise their kits' vocalisations? *Jens Malmkvist*	The behavioural and physiological responses of horses to routine handling. *Jenna Kiddie*
14.45	Behavioural indicators for cats' perception of food: A novel tool for development of veterinary pharmaceuticals. *Helena Telkanranta*	Tourism and marine mammal welfare. *Donald Broom*

15.00	Break and poster session 1 ....
16.00	ISAE AGM
17.30	**Reception at the Town Hall of Aarhus (1 h)**

Wednesday 9 August

9.00	**PLENARY: Mike Appleby:** **Understanding human and other animal behaviour: ethology, welfare and food policy.** **Venue: Aarhus Music Hall, Rhythmic Hall**

	Session 5, Topic 04 at Rhythmic Hall: Applying ethology in the keeping of animals	Session 6, Topic 02 at Small Hall: Free papers
9.30	The welfare of working dogs in cars. *Lena M. Skånberg*	The limits of animal welfare concepts. *Jes Harfeld*
9.45	Do cull dairy cows become lame from being transported? *Kirstin Dahl-Pedersen*	Behaviour and brain gene expression of environmentally enriched pigs. *Alistair Lawrence*
10.00	The effect of overstocking different resources within a freestall pen on the behaviour of lactating Holstein cows. *Clay Kesterson*	Can body nosing in early-weaned piglets be prevented by artificial sucking and nosing devices? *Roland Weber*

Wednesday 9 August

10.15	Effect of water temperature on drinking behavior and stress response of sheep. *Shin-ichiro Ogura*	Multiple enriched cages increase silver fox cubs' confidence and reduce stereotypic head twirls. *Anne Lene Hovland*

10.30   **Break and poster session 2** ....

	Session 7, Topic 04 at Rhythmic Hall: Applying ethology in the keeping of animals	Session 8, Topic 05 at Small Hall: Animal stress responses
11.30	Effects of genetic background and selection for egg production on food motivation in laying hens. *Anissa Dudde*	Who's stressed: nonlinear measures of heart rate variability may provide new clues for evaluating the swine stress response. *Christopher J. Byrd*
11.45	On-farm techniques used to minimize aggression in pigs by US pork producers. *Sarah H. Ison*	Effect of space allowance and simulated sea transport motion on the behaviour and heart rate of sheep. *Grisel Navarro*
12.00	If you give the pig a choice: suckling piglets eat more from a diverse diet. *Anouschka Middelkoop*	Effects of acute lying and sleep deprivation on behaviour and milk production of lactating Holstein dairy cows. *Jessie A. Kull*
12.15	Impact of straw particle size and reticulorumen health on the feed sorting behaviour of early lactation dairy cows. *Rachael Coon*	Can faecal cortisol metabolites be used as an Iceberg indicator in the on-farm welfare assessment system WelFur-Mink? *Anna F. Marsbøll*

12.30   **Lunch** ....

	Session 9, Topic 06 at Rhythmic Hall: Animal learning and cognition	Session 10, Topic 05 at Small Hall: Animal stress responses
13.30	From Pavlov to Bekhterev – and the pros and cons of positive and negative reinforcement. *Janne Winther Christensen*	The effect of a low dose of lipopolysaccharide (LPS) on physiology and behaviour of individually housed pigs. *Janicke Nordgreen*
14.00	Development of an ethical virtual fencing system for sheep. *Danila Marini*	Effects of embryonic norepinephrine on juvenile and mature quail behaviors. *Jasmine Mengers*
14.15	Motor self-regulation by goats (*Capra aegagrus hircus*) in a detour-reaching task. *Jan Langbein*	The influence of social stress on three selected lines of laying hens. *Patrick Birkl*
14.30	Cylinder size affects cat performance in the motor self-regulation task. *Katarzyna Bobrowicz*	Feather pecking: is it in the way hens cope with stress? *Jerine A. J. Van Der Eijk*
14.45	Factors affecting dairy calves' abilities to learn to use colour cues to locate a source of milk. *Jeffrey Rushen*	Do feather peckers, victims and control hens differ in symmetry of bilateral traits, body weight and comb size? *Fernanda M. Tahamtani*

15.00   **Break and poster session 3** ....

Wednesday 9 August

	Session 11, Topic 06 at Rhythmic Hall: Animal learning and cognition	Session 12, Topic 05 at Small Hall: Animal stress responses
16.00	Goats prefer happy human faces. *Christian Nawroth*	Investigation of response to the novel object test in pigs in relation to tail biting phenotypes. *Jen-Yun Chou*
16.15	Responses in novelty tests are associated with grain intake and unrewarded visits to the milk feeder in dairy calves. *Heather W. Neave*	Effects of elevated platforms on fearfulness in fast-growing broilers. *Ida J. Pedersen*
16.30	Light during incubation and noise around hatching affect cognitive bias in laying hens. *Bas T. Rodenburg*	Meloxicam and temperament effects on pain sensitivity and inflammatory response in surgical or rubber ring castrated calves. *Désirée Soares*
16.45	Facial expressions of emotion influence sheep learning in a visual discrimination task. *Lucille Bellegarde*	Cat responses to handling: assessment of scruffing, clips, and full body restraint. *Carly Moody*

18.30	**Conference dinner. Venue: Turbine Halls, Aarhus**

Thursday 10 August

9.00	**PLENARY: Jean-Loup Rault:** **Positive social behaviour: Inside out.** **Venue: Aarhus Music Hall, Rhythmic Hall**

	Session 13, Topic 07 at Rhythmic Hall: Social behaviour of animals	Session 14, Topic 04 at Small Hall: Applying ethology in the keeping of animals
9.45	Artificial intelligence algorithms as modern tool for behavioural analysis in dairy cows. *Oleksiy Guzhva*	Lowered tail can predict a tail biting outbreak in grower pigs. *Helle Pelant Lahrmann*
10.00	Best friends forever, new approach to automatic registration of changes in social networking in dairy cows. *Per Peetz Nielsen*	Using hen-mounted light sensors to monitor outdoor range use. *Stephanie Buijs*
10.15	Phenolab: ultra-wide band tracking shows feather pecking hens spent less time in close proximity compared to controls. *Elske N. De Haas*	Housing laboratory mice from the mouse's point of view. *Elin M. Weber*

10.30	**Break and poster session 4** ....

Thursday 10 August

	Session 15, Topic 07 at Rhythmic Hall: Social behaviour of animals	**Session 16, Topic 04 at Small Hall:** Applying ethology in the keeping of animals

11.30	Floor feeding over multiple drops does not increase feeding opportunities for submissive sows. *Megan Verdon*	Welfare relevance of floor area for single-housed female mink (*Neovison vison*). *María Díez-León*
11.45	The effect of socialization before weaning on social behaviour and performance in piglets (*Sus scrofa*) at a commercial farm. *Laura Salazar*	Differential effects of enrichment on the subtypes of stereotypic behaviour in mink. *Andrea Polanco*
12.00	Socialisation, play behaviour, and the development of aggression in domestic pigs (*Sus scrofa*). *Jennifer Weller*	Effects of space allowance and simulated sea transport motion on the behavior of sheep. *Ramazan Col*
12.15	Intensity of aggression in pigs depends on their age and experience at testing. *Irene Camerlink*	Cooling cows with soakers: effects of flow rate and spray frequency on behaviour and physiology. *Grazyne Tresoldi*

12.30   Lunch . . . .

	Session 17, Topic 08 at Rhythmic Hall: Animal affective states	**Session 18, Topic 09 at Small Hall:** Maternal and neonatal behaviour

13.30	What can facial expression reveal about animal welfare? Supporting evidence and potential pitfalls. *Kris Descovich*	The affective dyad: the role of maternal behaviour in positive affective states of mother and young. *Cathy M. Dwyer*
14.00	Do dogs smile when happy? An objective and comparative study of dog and human facial actions in response to emotional trigger. *Cátia Correia-Caeiro*	Dairy cows and heifers prefer to calve in a bedded pack barn or natural forage compared to open pasture. *Erika M. Edwards*
14.15	Facial expression of harbour seal pups (*Phoca vitulina*) in response to painful stimuli. *Amelia Mari Macrae*	Factors affecting maternal protective behavior in Nellore cows. *Marcia Del Campo*
14.30	The startle response as a measure of fear in dairy calves. *Sarah J. J. Adcock*	Preference for mother does not last long after weaning in lambs. *Rodolfo Ungerfeld*
14.45	Pre-weaning environment affects pigs' emotional reactivity. *Oceane Schmitt*	Behavioral responses of dairy calves to separation: the effect of nutritional dependency on the dam. *Julie Føske Johnsen*

15.00   **Break and poster session 5** . . . .

	Session 19, Topic 08 at Rhythmic Hall: Animal affective states	**Session 20, Topic 09 at Small Hall:** Maternal and neonatal behaviour

Thursday 10 August

16.00	Lying around with nothing to do: boredom in farmed mink. *Rebecca K. Meagher*	Does farrowing duration affect maternal behaviour of outdoor kept sows? *Cecilie Kobek Thorsen*
16.15	Can minor, easily applied alterations of routines during the rearing period reduce fearfulness in adult laying hens? *Margrethe Brantsæter*	Does dietary tryptophan around farrowing affect sow behavior and piglet mortality? *Jeremy N. Marchant-Forde*
16.30	Assessing anxious states in sheep: a more practical attention bias test. *Jessica Monk*	Relationships between sow conformation, accelerometer data and crushing events in commercial piglet production. *Stephanie Matheson*
16.45	Development of behavioural tests for WelFur on-farm welfare assessment of foxes. *Jaakko Mononen*	Effects of litter size on maternal investment and neonatal competition in pigs with different genetic selection pressures. *Marko Ocepek*

17.10   **Closing of conference. Venue: Aarhus Music Hall, Rhythmic Hall**

18.00   **Farewell Reception. Venue: Aarhus Music Hall**

Table of contents

Topic 02. Free papers

Topic 03. Human-animal interactions

Oral presentations – Session 4

Poster presentations – poster session 1

Topic 04. Applying ethology in the keeping of animals

Oral presentations – Session 5

Topic 05. Animal stress responses

Oral presentations – Session 8

Oral presentations – Session 10

Topic 06. Animal learning and cognition

Oral presentations – Session 9

Oral presentations – Session 11

Poster presentations – poster session 3

Topic 07. Social behaviour of animals

Oral presentations – Session 13

Artificial intelligence algorithms as modern tool for behavioural analysis in dairy cows 174
Håkan Ardö, Oleksiy Guzhva, Mikael Nilsson, Anders Herlin, Lena Lidfors and
Christer Bergsten

Best friends forever, new approach to automatic registration of changes in social
networking in dairy cows 175
Per Peetz Nielsen, Luis E.C. Rocha, Olle Terenius, Bruno Meunier and Isabelle Veissier

Phenolab: ultra-wide band tracking shows feather pecking hens spent less time
in close proximity compared to controls 176
Elske N. De Haas, Jerine A.J. Van Der Eijk, Bram Van Mil and T. Bas Rodenburg

Oral presentations – Session 15

Floor feeding over multiple drops does not increase feeding opportunities for
submissive sows. 177
Natalia Zegarra, Paul. H. Hemsworth and Megan Verdon

The effect of socialization before weaning on social behaviour and performance
in piglets (*Sus scrofa*) at a commercial farm 178
Laura Salazar, Irene Camerlink, Heng-Lun Ko, Chung-Hsuan Yang and Pol Llonch

Socialisation, play behaviour, and the development of aggression in domestic
pigs (*Sus scrofa*) 179
Jennifer Weller, Simon Turner, Irene Camerlink, Marianne Farish and Gareth Arnott

Intensity of aggression in pigs depends on their age and experience at testing 180
Irene Camerlink, Marianne Farish, Gareth Arnott and Simon P. Turner

Poster presentations – poster session 4

Effect of overstocking during the dry period on the behaviour and stress level of
dairy cows, and on pre-weaning calf growth 181
Mayumi Fujiwara, Kenny Rutherford, Marie Haskell and Alastair Macrae

Relationship between social rank and reproductive, body and antler traits of
pampas deer (*Ozotoceros bezoarticus*) males 182
Matías Villagrán, Florencia Beracochea, Luděk Bartoš and Rodolfo Ungerfeld

Rearing bucks (*Capra hircus*) isolated from females affects negatively their sexual
behavior when adults 183
Lorena Lacuesta, Julia Giriboni, Agustín Orihuela and Rodolfo Ungerfeld

Topic 08. Animal affective states

Oral presentations – Session 17

Topic 09. Maternal and neonatal behaviour

Oral presentations – Session 18

Oral presentations – Session 20

Poster presentations – poster session 5

Wood-Gush Memorial Lecture:
Echolocating whales as models for studying how information flow governs behavior
Peter Teglberg Madsen
Aarhus University, Department of Bioscience, C.F. Møllers Allé 3, 8000 Aarhus C, Denmark;
peter.madsen@bios.au.dk

Sensible behavioral responses of an animal to the environment depend critically on the capability to sample its surroundings with a suite of sensory systems. Quantification of the sensory flow available to animals in time and space is therefore crucial for understanding behaviors in the wild such as predator avoidance, foraging strategies and mate selection. However, animals are exposed to a barrage of information in their natural habitats, and it is inherently difficult to quantify which sensory cues are extracted by their sensory systems to inform behavioral transitions. Further, because most animals perceive their surroundings on the basis of multimodal sensory integration, behavioral changes are often precipitated by a complex set of sensory inputs making it difficult to isolate a single causal cue. Moreover, sensing is an interactive process in which animals modulate their information flow by continuously adjusting their attention and sensor performance by changing sensitivity and directivity of each sensory modality. Keeping track of all of the potential sensory inputs that may guide behavior is therefore a major experimental challenge even in a controlled laboratory setting let alone in a free-ranging animal in the wild: there is inevitably a trade-off between how well sensory inputs and attention can be measured, and the impact of the measurement apparatus itself on the animal. With this talk I wish to introduce echolocating toothed whales as models for overcoming at least some of these challenges. Echolocation is a process where the animals must emit sound to generate echoes returning to the auditory system for processing, meaning that they control sensory information flow by the rate, type and direction of the sounds they produce as well as by adjusting the sensitivity of their hearing. These parameters directly influence the temporal resolution and spatial extent of their perception of their environment, enabling dynamic control of attention in response to environmental complexity and behavioral objectives. Thus, the way that echolocating toothed whales manipulate their perception of the surrounding environment is revealed by the sonar pulses they emit, the echoes they receive and the evoked behavioral changes; all of which can now be sampled by small archival tags attached to their backs with suction cups for up to 3 days. I will argue that echolocating toothed whales are ideal experimental models for studying sensory and cognitive adaptations that confer advantages in handling a dynamic sensory umwelt in time and space, providing a unique, non-invasive insight into perception in a naturally behaving animal. I hope that such examples from using multisensor tags on echolocating whales will inspire applied ethologists working with farm, laboratory and companion animals to further use biologging technology to quantify the relationships between complex stimuli and behavior.

Understanding human and other animal behaviour: ethology, welfare and food policy

Michael C. Appleby
University of Edinburgh, Easter Bush, Edinburgh, EH25 9RG, United Kingdom;
michael.appleby@ed.ac.uk

Food security and sustainability are paramount goals of global policy in the Twenty First Century. What are the contributions of applied ethology and animal welfare in delivering these major goals? Livestock plays a major role in food security and nutrition (FSN), because meat and animal products are important in people's diets and the livestock sector is central to food systems' development. And many aspects of the welfare of livestock are important for production, as expressed in the advice, 'Look after your animals and they will look after you,' for example, reduction of disease. Consideration of animal welfare can enable advantages for production that have not otherwise been identified, and is therefore not just compatible with but important for achieving good management of livestock for production, livelihoods and FSN. Furthermore, understanding of animal behaviour and care for livestock welfare contribute to all three pillars of sustainable agriculture: economic profitability, social equity and environmental health. Ethology clarifies the interactions between animals, humans and the environment, and thus helps to optimise livestock management for the best balance between these complex priorities. However, although considering behaviour and welfare is valuable for production and FSN, people do not always recognise or act on this value. They may need information or help to do so: mechanisms are needed to provide information, help and sometimes financial assistance to producers. These mechanisms need input from experts in applied ethology and animal welfare, at both practical and policy level. Six policy areas will be briefly outlined. Each requires (to a variable degree) understanding of animal behaviour and welfare and implementation of that understanding. For example, (1) Development of humane, sustainable food security strategies, requires implementation of animal welfare standards, research and development into practical improvement of management and welfare on farms, and knowledge transfer to farmers. The other areas are (2) Decisions on intensification of livestock farming; (3) Feeding of arable crops, especially cereals, to livestock; (4) Development of specific food and livestock policies to assist vulnerable sectors of the population; (5) Sustainable diets; and (6) Development of markets for humane, sustainable livestock production. Increased urgency is needed to achieve food security and sustainability, by producers, governments and intergovernmental organisations in consultation with other stakeholders including scientists and non-governmental organisations. The livestock sector has both positive and negative impacts. Consideration of animal behaviour and welfare helps to ensure and increase the positive impacts, and to reduce and avoid the negative, and should therefore be a priority for countries worldwide.

Positive social behaviour: inside out

Jean-Loup Rault

University of Melbourne, Animal Welfare Science Centre-Faculty of Veterinary and Agricultural Sciences, Alice Hoy building 162, Parkville, 3010, Australia; raultj@unimelb.edu.au

Social behaviour is central to the life and well-being of humans and domestic animals. Nevertheless, we know much less about positive social behaviours (the 'bright side' of living in a group) in comparison to negative social behaviours (the 'dark side', albeit necessary). The definition of positive social behaviour itself has challenged psychologists for decades. As an operational definition, a positive social behaviour (also referred to as prosocial behaviour, hereafter) refers to an action that an individual engages in intended to benefit others. Examples include affiliative (e.g. allo-grooming, spatial proximity, behavioural synchronisation), maternal, cooperative and other types of caring and helping behaviours toward another individual. Research on prosocial behaviours has been hindered by a poor understanding of their mechanisms. Evidence from complementary fields such as behavioural neuroscience provides crucial insight for applied behaviour science; helping us understand how a behaviour is perceived as prosocial and how it affects the recipient. The oxytocin family hormones play a crucial role in social communication and social cognition, both essential conditions for prosocial interactions. Whether oxytocin's action is specific to prosocial behaviours is still debated. Behavioural science itself has much to contribute: discrepancy in the literature on prosocial behaviours may be partly explained by their modulation by contextual features (i.e. features of the situation in which prosocial behaviours are shown, or conversely not shown); a parameter that has often been overlooked. Understanding the role of ontogeny and identifying means to encourage prosocial behaviours are particularly relevant to improve the way we manage and care for captive social animals. However, important questions remain to be addressed. For instance, which situations encourage prosocial behaviours, group stability vs conflict (the 'social buffering' effect)? Furthermore, are we overlooking benefits the emitter may receive from delivering prosocial behaviours, and is reciprocity important? Also, what effects prior experience, and particularly ontogeny, has on the ability to process and benefit from prosocial behaviours? This knowledge is key to implement solutions that promote positive experiences, given the powerful influence of social factors on well-being. Prosocial behaviours also cross species boundaries, as illustrated by the growing number of studies on the effects of positive human-animal relationships. Ultimately, the close relationship between social behaviour and emotions offers tremendous opportunities to utilise prosocial behaviours in practice for promoting positive mental states and well-being.

Key-note:
The role of olfaction in relation to stress and fear
Vincent Bombail
INRA – Institut National de la Recherche Agronomique, Neurobiologie de l'Olfaction, Domaine de Vilvert, 78350 Jouy en Josas, France; vincent.bombail@inra.fr

In a world populated with more competitors and predators than friends, sensory information can guide and help animals adapt to their environment. Olfaction is a key sensory modality in several animal species. It affects many behaviours related to essential functions such as feeding, reproduction or social interactions. The ability to detect and recognise odorant molecules is affected by the inner state of the animal, indeed olfactory detection can undergo changes in order to adapt perception to physiological needs. In turn, odours can include alarm signals that induce fear or a stress response. The stress response is an important mediator of the adaptation process, linking physiology and behaviour in response to environmental challenges. In this overview of the literature, we will reflect on the interplay between stress or fear and olfaction, with a view to describe recent discoveries relevant to the study of animal behaviour, and to highlight interesting gaps in our knowledge that might require further investigation (e.g. how to assess behavioural alterations caused by altered senses?). First, experimental evidence will be presented that supports an effect of the stress response on olfactory detection. Acutely or chronically stressed animals might indeed perceive odours differently (quantitatively and qualitatively); we will discuss how this might affect animal behaviour. Next we will review the abundant literature on odours that can cause fear or stress in animals. Predator odours and molecules purified from predator bodily secretions are well documented inducers of fear and the stress response. Interestingly, certain odours from stressed conspecifics can also elicit physiological and behavioural signs of alarm in naive individuals. Examples of this chemical communication and its impact on animal behaviour and welfare will be discussed. Finally, we review the claim that odours can alleviate the effects and manifestations of the stress response. This intervention seems an elegant way to improve animal welfare, indeed several researchers report that exposure to certain odorant compounds, such as essential oils, can have anxiolytic or antidepressant effects. Concepts and data from seemingly unrelated fields of neuroscience such as olfaction and stress biology should be taken into account in experimental design and interpretation, for a better understanding of animal behaviour. Perhaps, as a species that do not often use olfaction as our primary sensory modality, we humans might lack some empathy towards animals that do. In order to improve the welfare of captive animals, we should be more aware of these aspects of animal perception.

Olfactory enrichment for cattle: development of tests for determining olfactory investigation of complex odours in cattle

Maria Vilain Rørvang[1], Margit Bak Jensen[1] and Birte Lindstrøm Nielsen[2]
[1]Aarhus University, Department of Animal Science, Blichers Allé 20, 8830 Tjele, Denmark,
[2]INRA, NeuroBiologie de l'Olfaction, Université Paris-Saclay, 78350 Jouy-en-Josas, France; maria.
vilainrorvang@anis.au.dk

Olfaction is the main sensory modality in the majority of mammalian species, playing a key role in their interactions with the environment. However, the link between chemical signals and animal behaviour has often been ignored, especially with respect to large domestic species. In cattle, there may be unexploited potential for using odours and olfaction in their management. By applying an olfactory habituation/dishabituation test originally developed for rodents, this study aimed to assess olfactory ability in cattle. Group housed, pregnant, dry cows (n=10) and heifers (n=13) were presented with three different odours (orange juice, liquid coffee and a tap water control placed in a test bucket) in randomised order. The test was conducted on individual animals in their home pen and consisted of each odour being presented twice in a row for 2 min each with an inter-trial pause of 2 min. Following another 2-min pause without odour the animal was presented with the new odour, with order of odour presentation balanced between animals. Duration of sniffing (muzzle in proximity or contact with) the test bucket as well as the occurrence of licking or biting the test bucket were recorded by direct observation. All animals sniffed an odour (i.e. the test bucket) less when presented for the second time (habituation; Median (IQR): 3.01 (0.43-11.61), 0 (0-5.14), 15.19 (4.73-42.50), 3.04 (0-12.08), 8.61 (2.72-46.76) and 1.44 (0-7.69) for Water1, Water2, Coffee1; Coffee2, Orange1 and Orange2, Wilcoxon-Pratt signed-rank test: all 1st vs 2nd presentations: z^3 4.20, P<0.001). Sniffing duration increased after presentation of a new odour (dishabituation; all 2nd vs 1st presentations: z £ -3.74, P<0.001). All animals sniffed coffee and orange longer than water (water vs coffee, z=-4.17, P<0.001 and water vs orange, z=-3.74, P<0.001), but the animals also sniffed coffee longer than orange (z=2.28, P=0.021). Licking or biting behaviour occurred only when presented with coffee or orange (13 out of 23 animals for both samples). No effect of parity was found for sniffing, licking or biting behaviour. This is the first example of an olfactory test being adapted for and applied to cattle, which adds to our knowledge of what types of odours bovines are capable of smelling. The test showed that cattle can distinguish between different complex odours (coffee and orange), and that increased interest was evoked for one of these odours (coffee). These results are the first steps towards exploring the possibility of using odours when adapting or enriching the environment in which we keep cattle and improving the management practices currently applied.

Do dogs see the Ponzo illusion?

Sarah-Elizabeth Byosiere, Lynna C. Feng, Nicholas J. Rutter, Jessica K. Woodhead, Philippe A. Chouinard, Tiffani J. Howell and Pauleen C. Bennett
La Trobe University, Psychology and Public Health, P.O. Box 199, Bendigo Victoria 3552, Australia; s.byosiere@latrobe.edu.au

Domestic dogs play many roles in human lives, but little is known about how they perceive their environment. While dogs possess an exceptional olfactory sense, dog cognition is commonly assessed via visual tasks. This makes it important to understand how dogs perceive simple visual stimuli. One way to begin to uncover visual capabilities in animals is to assess perception of geometric illusions, in which mechanisms that are normally helpful for accurately perceiving the environment trick the brain into misperceiving certain visual stimuli. Using a two-choice discrimination paradigm, eight dogs were trained to select, by touching with their nose, either the larger (7 dogs) or smaller (1 dog) of two circle stimuli presented on a screen. They then underwent familiarization sessions in which the circle stimuli were embedded in non-illusory backgrounds, before being assessed for their susceptibility to various presentations of the Ponzo illusion across four separate experiments. Test sessions comprised control presentations, where stimuli differed in size, and illusion presentations, where stimuli were identical in size. Correct choices in the illusion condition were defined as selecting the stimulus typically chosen by susceptible humans in similar tasks. Two-tailed binomial tests assessed each dog's responses in both conditions to determine if they were performing significantly above chance. Effect sizes were calculated for each illusion condition based on the means of the group. To examine group performances all illusions were analysed using two-tailed one-sample t-tests conducted on the dogs' average percent correct. A one-sample t-test of dogs' responses in the first experiment demonstrated illusion susceptibility at the group level [$t(7)$=2.65, P=0.033, SD=6.68, Cohen's d=0.9328]; however, no individual dog performed significantly above chance in binomial tests (P$\geq$0.06 for all tests). In three subsequent experiments one-sample t-tests found no significant results at the group level [$t(7)$=0.42, P=0.685, SD=12.52, Cohen's d=-0.1498; $t(5)$=0.66, P=0.53, SD=0.15, Cohen's d=-0.269; $t(6)$=0.95, P=0.37, SD=18.64, Cohen's d=0.34, respectively]. Only one dog demonstrated a small but significant tendency towards susceptibility (P<0.03). Taken together, then, there was no evidence for dogs' susceptibility to the Ponzo illusion in a two-choice discrimination paradigm. As most animals tested previously have demonstrated susceptibility, these findings have implications for theoretical explanations of the Ponzo illusion, which, in humans, typically posit a reliance on high levels of visual processing. The divergence of results between dogs and humans/other animals suggest that mechanisms underlying perception of the Ponzo illusion may differ across species. They also suggest that care should be taken when using visual paradigms to test dogs' cognitive skills.

Key-note:

Taking into account different sensory abilities of animals in their housing and management

Birte L. Nielsen
INRA, Neurobiologie de l'Olfaction, Université Paris-Saclay, 78359 Jouy en Josas, France;
birte.nielsen@inra.fr

Animals as well as humans interact with the surroundings in a way, which reflects their ability to receive and process information. As humans, most of us rely heavily on our eyes and ears when we go about our daily lives, whether it is looking out for passing cars or listening to music. When it comes to the sense of smell, however, we do not think it is as important as vision and hearing. Animals differ from us in many ways when it comes to use of sensory modalities. Some species have sensory abilities we can only dream of, such as the echolocation of bats and certain birds, and the electric field sensing of sharks. Animals also use the sensory modalities we do share to a different extent than us, and often with a higher acuity. The olfactory ability of rats, for example, is much more developed than our sense of smell; the rat being a nocturnal animal, which in nature rarely ventures out during the day. Likewise, birds of prey have a well-developed visual ability, being able from a great height to spot and track tiny prey on the ground. So what are the implications of this for the housing, management, and welfare of animals in our care? In essence: that we need to try and sense the world from the animal's point of view. This will depend on the species in question and the situation we are trying to manage. Moving cattle through a corridor with minimum stress obviously demands different procedures than those needed to provide a laboratory rat with suitable housing. As scientists, as farmers, as pet-owners, and as zoo-keepers, we need to know and understand the sensory abilities of the species we deal with. In the presentation, I will discuss how differences in sensory abilities between species affect the behaviour and welfare of animals managed by humans. This will be based on illustrative examples from mammals and birds, and touch upon issues such as lighting, enrichment, and handling. Different sensory modalities will be covered, and for once I will refrain from putting special emphasis on the importance of olfaction.

The impact of ultraviolet wavelengths on broiler chicken welfare indicators

Charlotte James[1], Lucy Asher[2] and Julian Wiseman[1]
[1]University of Nottingham, Animal Science, Sutton Bonington Campus, Sutton Bonington, Leicestershire, LE12 5RD, United Kingdom, [2]Newcastle University, Centre for Behaviour and Evolution IoN, Henry Wellcome Building, Framlington Place, Newcastle, NE2 4HH, United Kingdom; charlotte.james@nottingham.ac.uk

The aim of the study was to investigate the impacts of Ultraviolet wavelengths (UV) on a variety of welfare indicators in broiler chickens. We explored UVA and UVB wavelengths. UVA is visible to chickens and may provide a valuable source of environmental enrichment, in turn facilitating more harmonious flock interactions. UVB wavelengths promote endogenous vitamin D synthesis, which could support the rapid skeletal development of broiler chickens. Day-old Ross 308 birds were randomly assigned to one of three treatments; C) White Light Emitting Diode (LED) control group, representative of farm conditions (18-hour photoperiod). A) White LED & supplementary UVA LED lighting (18-hour photoperiod) and AB) White LED (18-hour photoperiod) with supplementary UVA & UVB fluorescent lighting providing 30 micro watts/cm^2 UVB at bird level (on for 8 hours of the total photoperiod to avoid over-exposure of UVB). Birds were kept at a final stocking density of 33 kg/m^2 and fed ad-lib on a commercial diet. Welfare indicators measured were; feather condition (day 24, n=546), tonic immobility duration, a measure of stress responsiveness (day 29, n=308), and gait quality, using the Bristol Gait Score (day 31, n=299). For statistical analysis, generalised linear or ordinal logistic regression models were fitted in R statistical software. Results showed, improved feather condition in treatment A, compared the control (P=0.03). Treatment A birds also had a lower average tonic immobility duration of 1.02 (SE 0.10) minutes, compared to 1.52 (SE 0.12) minutes in the control group (P=0.03). Lighting condition had no effect on Bristol Gait Score. Results suggest UVA may be beneficial for broiler chicken welfare. While treatment A and AB both provided UVA, the improvements in welfare indicators were not consistent. One possible explanation is that the beneficial effects of UVA are exposure time dependent. Future research identifying the links between UVA exposure and positive impacts on feathering rate, stress, activity patterns, spatial distribution and flock interactions will be of further importance to broiler welfare.

How incline angle affects the preparatory step prior to climbing a ramp in domestic fowl

Chantal Leblanc[1], Bret Tobalske[2], Bill Szkotnicki[1] and Alexandra Harlander[1]
[1]University of Guelph, Animal Biosciences, 50 Stone Road East, Guelph, Ontario, N1G 2W1, Canada, [2]University of Montana, Division of Biological Sciences, 32 Campus Drive, Missoula, MT, 59812, USA; aharland@uoguelph.ca

For laying hens, daily activities require that they adapt their walking pattern to account for different challenges of their environment. Especially in three-dimensional rearing systems, birds have much more space and opportunity to interact with their environment. To help them safely navigate this complex setting, ramps and ladders are provided for reaching high-elevation tiers of the housing structures. Perceiving these obstacles and modifying basic locomotion patterns in an anticipatory manner to safely navigate different surfaces is referred to as adaptive locomotion. Individuals visually perceive the environmental properties or obstacles ahead of them, and modify their locomotion patterns in an anticipatory manner. We measured the effects of incline angle upon the locomotor adjustments of the preparatory step prior to climbing a ramp. We predicted that the magnitude of kinetic changes would be greatest for steep inclines, specifically, that birds would modulate kinetics to maximize peak ground reaction force (GRF), and maximize ground contact time. We assessed the GRFs using a walkway integrated force plate, collecting force data one step prior to climbing the ramp. We recorded 20 female domestic fowl preparing to ascend ramp inclines at slopes of +0, +40 and +70° when birds were 17, 21, 26, 31, and 36 weeks of age. Birds had experience with the ramp apparatus over a series of sessions before testing. Each bird was tested once per treatment. Hens were provided with motivation to ascend using 5 pen mates in a crate in addition to providing a table spoon of their feed along with raisins. We tested for effects of incline upon locomotion kinetics using PROC GLIMMIX in SAS (9.4). We conducted a repeaed measures variance analysis for each variable. Age of the birds, incline and interaction (age×incline) were included as fixed effects. There was no effect of age on kinematic force plate data. We found that GRFs of the foot changed gradually for increasing slope angle. There were significantly higher vertical peak GRFs and longer level ground contact times in preparation for the steepest inclines. The magnitude of the peak vertical GRFs (relative to bodyweight) showed a significant increase at the greatest angle (F_2, 37.55=28.88, $P<0.0001$). There were no significant differences between horizontal and +40° ramp inclines, whereas higher vertical peak GRFs were measured on +70° ramp inclines compared to +40° ramp incline (-1.31±0.18, $t17.49$=-7.32, $P<0.0001$) and compared to the horizontal ramp (-1.34±0.18, $t17.08$=-7.58, $P<0.0001$). Contact time depended on the angle of the incline (F_2,41.28=10.73, $P=0.0002$). Our results reveal that domestic fowl modulate their locomotion in response to incline angle, possibly in an anticipatory manner. This information about bipedal kinetic modulation is essential for our understanding of the control for locomotion in laying hens.

Do mink mothers recognise their kits' vocalisations?

Jens Malmkvist
Aarhus University, Animal Science, Blichers Alle 20, 8830 Tjele, Denmark;
jens.malmkvist@anis.au.dk

Offspring recognition requires fulfilment of at least two fundamental criteria: (1) offspring must produce distinctive signals and (2) parents must be able to discriminate between these signals. We have previously suggested that the vocalisations of offspring from farm mink (*Neovison vison*) meet the first criteria with reports of rich signature in kit calls already early in life. Mink kits produce complex ultrasonic vocalisations with energy up to 50 kHz with individual variation in e.g. pulse duration, frequency modulation rate and relative energy of harmonics. Likewise, the hearing ability of the mink dam is well developed with a broad hearing range from 1 kHz up to above 70 kHz. On farms, several mink are individually housed in the same facility, delivering around the same time of year, and farmers often transfer kits between litters of the same age within the first week of life. Levelling out of large litters is believed to increase kit survival. However, it is unknown whether mink mothers are able to differentiate between own vs unfamiliar kits' vocalisations. Therefore, we studied mink mothers' (n=18 second-parity brown, litter size 6 to 9) behaviour towards playbacks of own vs unfamiliar kit vocalisations day 2 postpartum. Each dam was randomly assigned to a block of playback sessions of offspring vocalisations recorded from individual male kits the previous day (postpartum day 1). The unfamiliar kit came from distant, not neighbouring cages relative to the test dam. The playback loudspeaker – with a high linearity output in the range 1-125 kHz – was placed just outside the end of the home cage opposite to the nest box opening. To reduce novel object exploration and to allow dams to habituate to the test in their home cage, a fake loudspeaker was placed in the playback position more than 24 h before the playbacks. Each dam was exposed to fixed orders of own and one unfamiliar male kit in call bouts of 1 min, controlled for order in the presentation for own and unfamiliar calls. Successive tests within session were separated by pauses of 5 min. Dam behaviour was recorded from video, and data analysed used Mixed model ANOVA taking repeated measures per dam into account. Dams reacted to the playback with approach, moving from the nest box with their litter into the wire cage closer to the source of the sound (time spent in wire cage: 69.1% with vocalisation, 14.2% when silent, P<0.001), also spending more time in contact with the loudspeaker (39.8%) during kit vocalisation than when silent (0.3%; P<0.001). The time spent in the cage was negatively associated with the litter size (P=0.035); litter size tended to decrease behavioural shifts (P=0.069), but not dam contact (P=0.12) during playback sessions. The dams explored (touched, sniffed) the back wall area with the loudspeaker for significantly longer time during the playback of her own kit (time spent: 19.0±2.11%) vs an unfamiliar kit (12.5±2.20%; P=0.040). This result and preference is the first evidence for mink mothers' recognition of own vs unfamiliar kits' calls.

Behavioural indicators for cats' perception of food: a novel tool for development of veterinary pharmaceuticals

Saila Savolainen[1], Helena Telkanranta[1,2], Jouni Junnila[3], Jaana Hautala[1], Sari Airaksinen[1,4], Anne Juppo[1], Marja Raekallio[1] and Outi Vainio[1]
[1]University of Helsinki, P.O. Box 57, 00014 University of Helsinki, Finland, [2]University of Bristol, Stock Lane, Langford, Bristol, BS40 5DU, United Kingdom, [3]4Pharma Ltd., Ahventie 4, 02170 Espoo, Finland, [4]Orion Pharma, P.O. Box 65, 02101 Espoo, Finland; helena.telkanranta@bristol.ac.uk

Administering oral medication to cats is often challenging to owners. Cats have a strong aversion to swallowing objects they perceive as non-food, and voluntary consumption of veterinary tablets by cats is typically below 50%. The pharmaceutical industry recognises the need to develop veterinary pharmaceuticals with a more acceptable smell, taste, size and consistency, but development is hindered by lack of research methods capable of detecting subtle differences in cats' perceptions. Such methods could help determine which formulations under development are less aversive than others, so that they can be selected for further improvement. Existing palatability tests are based on the less informative dichotomous measure of whether or not the cat eats a tablet; and on assessments by owners, subject to bias. Behavioural indicators offer a promising window to more detailed information on cats' perceptions. A set of behavioural indicators has previously been determined for two degrees of palatability of cat foods. The aim of this study was to identify an expanded set of behavioural indicators to measure subtle vs substantial degrees of palatability. A total of 34 pet cats were used in the study. The cats were presented with three types of edible items with a size of approx. 1 cm^3, selected individually for each cat according to preferences reported by the owner. FF (favoured food) was defined as a food preferred by the cat and having a consistency that makes it possible to hide a small tablet in it. TFF (tablet in favoured food) was the same food as above, with a placebo mini-tablet hidden in it. NFF (non-favoured food) was defined as something edible that the cat is unlikely to eat. The edible items were presented one at a time in a pseudo-randomized sequence. There were six trials per item, totalling 18 trials per cat, over a period of two days. Behaviour of cats before, during and after eating or refusing to eat was recorded on video. Two trained observers independently determined the prevalence of 16 behavioural patterns on the video recordings, blinded to the types of edible items. The data were analysed with a mixed logistic regression model. The following behavioural patterns were found to be more prevalent with NFF than with FF: a rapid ear flick backward (OR 12.6, P<0.001), a lick on the nose after not eating the item (OR 80.2, P<0.001), flicking of the tail (OR 7.3, P<0.001) and grooming of the body (OR 1.8, P<0.05). For the subtle difference between TFF and FF, the most promising indicator was the more prevalent dropping of food from the mouth while eating TFF (OR 2.0, P<0.001). These findings provide evidence of new behavioural indicators for objective assessment of food perception in cats, and they have practical applicability in designing a novel palatability test for developing veterinary pharmaceuticals with improved palatability for cats.

The effect of lighting provided during incubation on the initiation of feeding behaviour in broiler chicks

Miriam Gordon[1], Xuijie Li[1], Kayla Graham[1], Karen Schwean-Lardner[2] and Bruce Rathgeber[1]
[1]*Dalhousie University, Animal Science and Aquaculture, 58 River Rd, B2N 5E3 Truro, Canada,*
[2]*University of Saskwatchen, Animal Science and Poultry, 6D12 Agriculture Building 51 River Dr.,*
S7N 5A8 Saskatoon, Canada; miriam.gordon@dal.ca

Producing a high-quality broiler chick at hatch is important for subsequent growth and production of the bird. While commercial eggs are typically incubated in the dark, researchers have shown that *in ovo* photo-stimulation can influence embryo development and chicks may adapt more readily to the rearing environment. This study examined the initial placement behaviour of chicks that were incubated using different light spectrums. Ross 308 broiler eggs (n=1,600) were obtained from a commercial hatchery and randomly assigned to eight ChickMaster E09 single-stage incubators. Two incubators were operated with the traditional dark method of incubation (0L:24D), while the remaining 6 incubators were illuminated with white, red, or blue light (12L:12D). After 512 h of incubation, all hatched chicks were counted and gender determined. Upon arrival at the growing facility, chicks (n=1,248) were group weighed and randomly placed in 48 floor pens by block (incubator) with 26 birds of the same light colour treatment and gender in each pen (2.2×1.0 m). Chicks were all placed behind the waterline (0.49 m from back of pen). The time it took 50% of the birds to leave the back end of the pen and the time it took for 50% of the chicks to reach the feeder were measured immediately after placement. Broilers were raised under 18L:6D lighting condition. Lights came on at 05:00 h. On the day after placement, the number of chicks in the feed box were counted at 1-min intervals for the first hour after the lights came on. Average chick body weights and feed box weigh backs were measured 24 h after placement to corroborate feeding behaviour. The dark treatment chicks took the longest (160.8±13.5 sec) for 50% of the chicks to cross the waterline (P<0.05) and to have at least 50% of the chicks visit the feeder (221.0±27.2 sec; P<0.10) compared to white (99.0±13.5, 114.0±27.2), red (81.8±13.5, 119.0±27.2 sec), and blue (119.8±13.5, 160.5±27.2 sec). The average number of chicks counted in the feed box was least in the white light treatment (4.8±0.2) compared to red (6.2±0.2), blue (6.5±0.2), and dark (6.6±0.12; P<0.0001). The average weight gain (g/bird) and feed intake (g/bird) was greatest in the white light treatment (11.0±0.3 g; 8.2±0.2 g) compared to red (10.8±0.3 g; 7.93±0.2 g), blue (9.9±0.3 g; 7.8±0.2), and dark (9.6±0.3 g; 7.5±0.2 g; P<0.05). These results suggest that providing light during incubation can have positive influence on the behaviour of broiler chicks, which may subsequently improve their health and welfare. The effects of different spectrums of light are not clear.

Key-note:
Literature review on priority welfare issues of commercially raised bison in North America

Fiona C. Rioja-Lang[1], Jayson K. Galbraith[2], Robert B. McCorkell[3], Jeffrey M. Spooner[1] and John S. Church[4]
[1]*National Farm Animal Care Council, P.O. Box 5061, Lacombe, A.B., Canada,* [2]*Alberta Agriculture and Forestry, Livestock and Farm Business Section, 4910-52 St, Camrose, A.B., Canada,* [3]*University of Calgary, Faculty of Veterinary Medicine, 2500 University Dr. NW, A.B., Canada,* [4]*Thompson Rivers University, Dept. of Natural Resource Science, 900 McGill Road, Kamloops, B.C., Canada; jchurch@tru.ca*

A review of the scientific literature on bison was carried out as part of the National Farm Animal Care Council's (NFACC) process of updating Canada's current Code of Practice for the Care and Handling of Farmed Bison. Four priority welfare issues were identified including the effect of seasonality on the nutritional requirements of bison; bison behaviour; euthanasia on-farm; and pain. Bison reduce their feed intake and activity during winter, sparing energy reserves during periods of cold temperatures and food scarcity. To promote consistent carcass quality during this seasonality, producers commonly feed bison high grain diets. It has been suggested that feeding bison high concentrate finishing diets, greater than 80%, is likely to cause ruminal acidosis. Research is required into strategies that could help producers manage the effects of seasonality and to make it work to their advantage. Regarding behaviour, the handling of bison differs from traditional cattle handling techniques. Injuries and death during handling are more likely to occur if bison are handled incorrectly. They are more excitable than cattle, which have been bred for calm temperaments. Bison have a very intact social structure that has definite spacing requirements. The flight zone of bison tends to be much greater than that of cattle, and bison can be moved most effectively if the handlers work on the edge of it. Bison that become severely stressed during handling can suffer from a recognised condition known as capture myopathy. Handling facilities should restrict bison's vision and minimize loud noises. Bison should never be left in isolation as solitary bison display high levels of agitation. In the event of euthanizing bison on-farm, specific information regarding the gauge, calibre of firearm, and bullet selection is lacking. However, the combination of firearm and ammunition selected must achieve a muzzle energy of at least 1000 ft-lb (1,356 J) for animals larger than 400 lb. Understanding the correct landmarks for euthanizing bison are especially important as they are different from cattle. Due to the physical thickness of a bison skull, higher calibre firearms or heavier gauge shotguns are required than those used for other species. Bison producers in Canada rarely, castrate or brand animals, and they are seldom dehorned. There is currently a lack of peer reviewed research regarding the effect of painful procedures, however, there is little reason to suspect their physiological responses to pain are different to cattle. Both freeze branding and hot-iron branding cause pain and distress in bison, however in cattle, freeze branding causes less acute pain at the time of the procedure. The commercial bison industry is still relatively young, and continues to evolve. Expanding scientific studies will help to improve production whilst maintaining high welfare standards.

Associations between measures of rising and lying down behavior in dairy cows

Nina Dam Otten[1], Victor Henrique Silva De Oliveira[2], Mari Reiten[2], Anne Marie Michelsen[1], Franziska Hakansson[1], Vibe Pedersen Lund[1], Marlene Kirchner[1] and Tine Rousing[2]
[1]University of Copenhagen, Dept. of Veterinary and Animal Sciences, Grønnegårdsvej 8, 1870 Frederiksberg C, Denmark, [2]Aarhus University, Dept. of Animal Science, Blichers Alle 20, 8830 Tjele, Denmark; nio@sund.ku.dk

On farm welfare assessment should by international consensus primarily be based on an animal centred approach with a majority of outcome or so called animal-based measures. Although, these measures are regarded as having high construct validity they are also very time consuming to obtain and thus often replaced by resource-based measures or subject to a limited number of observations. In the Welfare Quality (WQ) protocol, a minimum of six observations per farm is regarded sufficient to assess lying down behaviour in cows. The objective of this cross-sectional study was to investigate associations between measures of rising and lying down behaviour in order to explore the possibility of measure reduction in future assessment protocols. Lying down behaviour including 'time needed to lie down' and 'collisions' was recorded in 60 Danish dairy herds according to the WQ protocol (min. six observations per herd) together with 'rising behaviour' among minimum 10% of the lactating cows. A total of 435 lying down and 1,332 induced rising sequences were evaluated. Mean time needed to lie down was assessed at herd level and rising sequences were categorized as 'normal' (pause on carpal joints less than three seconds, normal flow of movement), 'moderately problematic' (pause longer than three seconds and an otherwise normal rising sequence) or 'severely problematic' (pause longer than three seconds and other abnormal behaviour e.g. awkward head twisting, dog sitting, crawling backwards). Associations between the mean time needed to lie down and herd pervalences of observed collisions as well as mean herd level prevalences of the three rising behaviour categories were assessed in a linear regression model with herd as random effect and a backwards elimination at a 0.05 significance level. Arcsine transformation of the outcome and explanatory variables resulted in normal distribution of data. Overall mean time needed to lie down was 5.9 seconds (SD=3.7, median=5.2, min-max=1.8-51.3); collisions were present at an average of 33% (SD=34,median=25%, min-max=0-100) of the lying down sequences; herd level means for rising behaviour were: normal=78% (SD=17, median=82%, min-max=21-100), moderately problematic=17% (SD=14; median=14%, min-max=0-53) and severely problematic=5% (SD=8, median=2%, min-max=0-31.5). The final model showed significant associations between time needed to lie down and the percentage of collisions and severely problematic rising behaviour (P<0.001, R2=0.48). It can be concluded, that in order to reduce the number of measures and observations needed time needed to lie down might be sufficient to characterize the issues regarding cows' potential problems concerning both rising and lying down. Although, resource-based factors like cubicle design or stocking density are also potential influencing factors, these have not been included as the aim was to investigate relationships between the animal-based measures, rather than simply replacing these with resource-based measures.

Feedlot cattle are motivated to obtain roughage and show contrafreeloading

Jennifer Van Os[1,2], Erin Mintline[2], Trevor Devries[3] and Cassandra Tucker[2]
[1]University of British Columbia, Animal Welfare Program, BC, V6T 1Z4, Canada, [2]University of California, Department of Animal Science, Davis, 95616, USA, [3]University of Guelph, Department of Animal Biosciences, ON, N1G 2W1, Canada; cbtucker@ucdavis.edu

Feedlot cattle are commonly fed high-concentrate diets, which may lead to ruminal acidosis. In contrast, consuming roughage stimulates rumination and leads to more buffering of rumen pH. However, little is known about how motivated feedlot cattle are to consume roughage. A way to quantify motivation is requiring cattle to work (e.g. push a weighted gate) to access a resource. Using this method, we evaluated the motivation of feedlot cattle to obtain roughage when fed high-concentrate or high-fiber diets. During the study, and for 30 d before, 12 yearling Angus-Hereford cross heifers were fed either Sudan hay (high fiber, n=6) or a total mixed ration with 12% forage (as fed, high concentrate, n=6) *ad libitum*. These primary diets were delivered to individual cattle in an open feed bunk. Cattle were trained (≤4 d) to push and hold open a gate with their heads to obtain feed from a second bunk. Once data collection began, Sudan hay (200 g/d) was fed behind the gate, to which additional weight (34 kg/d) was added until heifers no longer pushed. Maximum weight pushed was expressed as % of bodyweight (BW). To compare consistently across days, behavioral data were evaluated on 5 d (d -4 to 0, relative to the final day they pushed the gate). All data were analyzed with generalized linear mixed models. We predicted cattle fed a concentrate-rich diet would push heavier weights, show a shorter latency, and spend more time using the gate, relative to those fed a high-fiber diet. Indeed, those fed a finishing diet pushed the gate immediately after hay delivery and much sooner than those fed a high-fiber diet (1.7 vs 75.7 min, respectively, back-transformed means, 95% confidence interval: 0.2-6.4 vs 40.0-131.0 min, P<0.001). This pattern was apparent even as the gate's weight increased: on the day before they no longer pushed the gate, cattle fed a high-concentrate diet showed a latency of only 2.9 min after hay delivery (vs 92.4 min for those fed a high-fiber diet). The suddenness with which heifers ceased pushing the next day suggests they were unable to move heavier weights to express their motivation. This may explain why maximum weight pushed (high concentrate vs high fiber: 46 vs 44% of BW, respectively, SEM: 3% of BW, P=0.73) and time spent using the gate (4.9 vs 5.8 min/d, SEM: 0.6 min/d, P=0.24) were similar between treatments. The small sample size may have limited our ability to detect treatment differences. Nonetheless, the considerable gate use by heifers with unrestricted access to hay is the first evidence in cattle of contrafreeloading (expending effort to obtain a resource even when it is simultaneously available freely), suggesting an internal motivation to work for food. In conclusion, we found evidence that consuming roughage is important to feedlot cattle, and this is affected by both their primary diet and an internal motivation to work to obtain feed.

The limits of animal welfare concepts

Jes Harfeld
Aalborg University, Center for Applied Philosophy, Kroghstræde 3, 9220, Denmark;
jlh@learning.aau.dk

Mink and pigs are some of the most intensively produced animals in Danish farming with an estimated production of 14 million pelts and 20 million pigs for slaughter per year. Taking this fact into account it is no wonder that Denmark has seen an increasing number of both public and academic debates about the welfare of animals in production systems. Much of the debate, however, revolves more around the possible legal and political consequences of animal welfare than around the concept of the welfare itself. Thus, the arguments from one side claims that animal welfare for pigs and mink is so poor that this should lead to the abolishment of the fur and pork production, while the other side claims that the welfare is good and that production of mink furs and pork loins should remain legal. It is an unresolved issue 'clouded in prejudice and misunderstandings on both sides'. It is not the main purpose of this article to take sides in this debate about legality. Instead, the concept and definition of mink and pig welfare itself is the turning point for this contribution and I will argue that much of the current debate is hindered by misconceptions about welfare. First of all, I will argue that the most important aspect of welfare – indeed what we actually mean by the concept – is an experiential phenomena, which traditional natural science, with its standard array of methods, has no tools to access directly. Marian S. Dawkins claims that much of current ethology too easily take the notion of animal experiential phenomena for granted and that it is problematic when we 'are asked to believe that the 'explanatory gap' between the physiology and behaviour we can observe and the subjective experience we cannot is no more than a small ditch that we can hop across'. I will argue that Dawkins is right, given her premises. I will, however, also argue that her premises, the claims about what science is, can and ought to be, are wrong, and that we can meaningfully address animal experiences within a scientific framework less rigid than her positivistic approach. Secondly, I will show how Dawkins' idea, although not taken entirely to heart, is acknowledged in the scientific literature by an – intentional or unintentional – narrowing of the concept of welfare. By attempting to approach welfare through an overly positivistic lens, the concept becomes removed from its fundamentally qualitative and experiential definition. Third and last, I will, inspired by philosophers Martha Nussbaum and Bernard E. Rollin, consider a more holistic concept of welfare and welfare assessment for mink and pigs based on the notion of species-typical thriving and flourishing. I will argue that such a holistic concept of welfare is necessary in order for us to gain a more adequate understanding of the animals and thus qualify the debate concerning their welfare.

Behaviour and brain gene expression of environmentally enriched pigs

Sarah Mills Brown[1], Rebecca Peters[2], Emma Baxter[2] and Alistair Burnett Lawrence[2]
[1]University of Edinburgh, Roslin Institute, Easter Bush Campus, Penicuik, EH25 6RG, United Kingdom, [2]SRUC, Roslin Institute, Easter Bush Campus, Penicuik, EH25 9RG, United Kingdom; alistair.lawrence@sruc.ac.uk

There is a clear need to provide further understanding of the biological relevance of environmental enrichment (EE) to pigs to promote EE in pig production systems. This project aimed to study the effects of EE on piglet behaviours and also for the first time on pig brain gene expression levels using RNA-sequencing technology. At 4 days post weaning, 8 piglets/ litter (from 6 litters) were selected as close to average litter weight as possible. Litters were split into 2 quartets balanced for sex, and housed in enriched (EE) or barren (B) pens. EE pens measured approx 3.6×1.8 m and were 2× the size of barren pens. Approximately 40 kg of long-straw was provided per EE pen to ensure the whole floor was well covered. Further enrichment was provided at the same time point daily to the 6 EE pens in the form of a plastic bag filled with straw (2 kg approx). Video footage was analysed using scan sampling from 15 min before and following bag provision (total=4 hours). On the final day 2 males per pen were euthanased under sedation via an intracardial overdose of pentobarbital sodium. Brain tissue was then collected and RNA extracted. Average active behaviours across scan samples were greater in the EE compared to the B pens ($t_9=3.04$, P=0.014). Differences in active behaviours (feeding, locomotion, object interaction, hops, pivots, nosing of penmate or pen) were most strongly observed at 1 hour post enrichment ($t_9=4.55$, P=0.001). At this time-point EE piglets were observed to interact with the bag in 19.5% of scans, with overall active behaviours occurring in 46.8% of scans (this compared with 22.6% in B piglets). RNA-sequence analysis of frontal cortex tissue showed EE relative to B piglets to have an over-representation of 9 gene ontology (GO) terms linked to biological processes with an under-representation of 14 biological process GO terms. Genes within over-represented GO terms in this dataset were those primarily associated with learning (e.g. *EGR2*) and neuronal signalling (e.g. *DUSP5*). Genes involved in feeding behaviour were also over-represented. These changes in brain gene expression are consistent with previous work showing increased learning capacity in EE pigs and with the trend towards higher average daily gain in the EE pigs in this study (F=3.21, P=0.087). Under-represented GO terms included a number involved in neuro-immunity (such as genes *IFI44L*, *IL1R1*, *CTSS*). As differential expression is determined as expression in one treatment relative to another, further work is required to understand if this reflects a more active neuro-immune response in B piglets or a down-regulation of neuro-immunity in EE piglets. Previous work in rodents has demonstrated lower expression of inflammatory cytokines in the hippocampus of EE mice compared to B mice when exposed to a virus. The present study demonstrates that environmental enrichment changes the expression of genes involved in learning and feeding behaviour and neuro-immune responses. These changes maybe linked to the level of active behaviours stimulated by EE housing.

Can body nosing in early-weaned piglets be prevented by artificial sucking and nosing devices?

Roland Weber[1], Daniela Frei[2], Hanno Würbel[3], Beat Wechsler[2] and Lorenz Gygax[2]
[1]Agroscope, Centre for Proper Housing of Ruminants and Pigs, Tänikon 1, 8356 Ettenhausen, Switzerland, [2]Federal Food Safety and Veterinary Office, Centre for Proper Housing of Ruminants and Pigs, Tänikon 1, 8356 Ettenhausen, Switzerland, [3]University of Bern, Division of Animal Welfare, Länggassstrasse 120, 3012 Bern, Switzerland; roland.weber@agroscope.admin.ch

In recent years, litter size has increased substantially by selection for hyper-prolific sows. Consequently, the number of live born piglets may exceed the number of functional teats, and surplus piglets are reared in artificial rearing systems as early as at the age of two days. In previous studies, it was found that early-weaned piglets show belly nosing and increased levels of other oral behaviours directed at pen mates. Belly nosing is considered as redirected suckling behaviour, as it resembles massaging and sucking at the udder. The development of such abnormal behaviour indicates that behavioural needs of the piglets are not met if raised without the sow. Moreover, belly-nosing may disturb resting behaviour of pen-mates and cause damage to the skin of their belly. The aim of this study was to examine whether body nosing (belly nosing plus nosing on any other part of the body) can be reduced or even prevented in piglets raised in an artificial rearing system. To do so, dummies designed to elicit sucking or massaging behaviour were presented to the piglets. During eight batches, a total of 144 Swiss Large White piglets were removed from the sow at two to five days of age and raised in groups of six piglets in artificial rearing units. A two-by-two between-group factorial design was applied in that a sucking and a massaging opportunity was either presented to the piglets or not. The behaviour of each piglet was scored from video for 90 min in six 15 min periods between 05:00 to 9:30 and 13:00 to 17:30 on days 4 and 18 after introduction to the artificial rearing unit. Data was analysed using linear-mixed effects models. The duration of body nosing increased in all treatment combinations between day 4 and day 18 from an average of 0.61 min (CI: 0.42-0.86) to 2.67 (1.87-3.84), but piglets raised in units equipped with a combined sucking/massaging dummy showed a much smaller increase from 0.37 (0.20-0.68) to 0.59 (0.35-1.03) min (three-way interaction: P=0.022). Piglets spent more time nosing the combined dummy (P=0.019) from an average of 0.17 (0.12-0.25) min than the dummies allowing for suckling and massaging only from 0.03 (0.02-0.04) min. The duration of body nosing was not directly related to the duration of dummy nosing (P=0.38). The time spent resting decreased in all treatments between day 4 with an average of 66 (63-69) min and day 18 with an average of 59 (57-62) min (P<0.001). Piglets with a combined dummy tended to a higher average resting bout length than piglets in other treatments (two-way interaction: P=0.066). To conclude, the combined dummy was most efficient in reducing body nosing. Additionally, piglets were less disturbed in their resting behaviour in this treatment. However, body nosing could not be prevented in any treatment. Therefore, the dummies tested do not seem to be sufficient in order to satisfy the piglets' behavioural needs for sucking and massaging.

Multiple enriched cages increase silver fox cubs' confidence and reduce stereotypic head twirls

Anne Lene Hovland
Norwegian University of Life Sciences, Department of Animal and Aquacultural Sciences, P.O. Box 5003, 1432 Ås, Norway; anne.hovland@nmbu.no

Access to environmental enrichment, defined as the addition of biological relevant objects and/or structures to captive animals' housing, is an important means to improve their welfare. Silver foxes kept for fur production are traditionally kept in wire mesh cages with access to a resting shelf and a gnawing object, and occasionally a nest box. Farmed foxes will eagerly explore their environment and show preferences for objects that can be chewed and manipulated and favour cages furnished with artificial dens (nest boxes) and shelves, devices found to support their welfare. Although foxes' have favourite objects, e.g. meat bones, access to several enrichments give multiple choices, potentially reinforcing activity and play. Here, we aimed to examine how access to a multiple enriched cage affected some welfare related parameters like fear towards humans, exploration, stereotypic head twirls and weight gain. In particular, knowledge on how enrichments affect the occurrence of abnormal behaviors in silver foxes is scarce and is thus important for improving our knowledge on captive fox welfare. Sixty 3-months-old cubs were pair housed where 15 pairs were housed in double standard cages additionally furnished with a tube, a wooden plate, a hockey-puck, straw and a meat bone. The controls had access to double standard housing conditions with a top nest box, a resting shelf and wooden gnawing sticks. To examine fearfulness and exploration a feeding test, a repeated titbit-test and a novel object test were conducted after 4 and 5 weeks. Stereotypic head twirls were recorded from video 30 minutes before feeding time after 6 weeks. The feeding test did not differentiate between the groups, but the titbit test revealed significant differences ($P \leq 0.05$) where more cubs from enriched cages accepted the titbit (53.3 vs 23.3%) after two trials. Immediately following handling and weight recording none of the controls accepted the titbit compared to 16.7% of the enriched foxes. Significantly more cubs from the enriched cages contacted the novel object (96.6 vs 70.0%; $P=0.012$).The latency to contact the object was shorter ($P \leq 0.003$) and the manipulation time was longer in the enriched group ($P=0.019$). The number of cubs with stereotypic head twirls were lower in the enriched group (33.3 vs 73.3%; $P=0.004$) and the level of head twirling also tended to be lower compared to the controls ($P=0.056$). After 5 weeks control cubs gained more weight compared to enriched cubs ($P=0.009$). Our results show significant effects of early access to multiple enrichments on cubs' behaviour, reducing fearfulness towards humans and novel objects, after only 6 weeks of access. Especially, the lowered incidence of head twirls indicates the potency of multiple enrichments to reduce frustrated motivations underlying the development of behavioural abnormalities. The results suggest addition of multiple enrichments as an important method for improving foxes' housing conditons. Future studies are needed to examine long term effects of early access to multiple enriched cages.

Use of unmanned aerial vehicles (UAVs) as a tool to measure animal welfare outcomes in beef calves

Justin Mufford[1], Desiree Soares[2], Karen Schwartzkopf-Genswein[2], David James Hill[1], Mark Paetkau[1], Nancy Flood[1] and John Church[1]
[1]Thompson Rivers University, 900 McGill Rd, V2B 0C8, Kamloops, BC, Canada, [2]Agriculture and Agri-Food Canada Lethbridge Research and Development Centre, 5403 1 Ave S, T1J 4P4, Lethbridge, AB, Canada; jt.mufford@gmail.com

Standard livestock management procedures (e.g. ear-tagging, branding, castration) can cause pain or stress to animals. To advance research on animal welfare, pain and stress mitigation strategies need to be better assessed under extensive field conditions. We developed a novel method to monitor cow-calf social affiliations by quantifying the proximity between animals using unmanned aerial vehicles (UAVs). Using two UAVs equipped with video-capable digital cameras, as well as photogrammetry and image analysis software, we measured the distance between cow-calf pairs. We hypothesized that this distance can be accurately measured by UAVs and has the potential to provide valuable information for studies of animal behavior and welfare. Twelve cow-calf pairs were uniquely identified with the use of penning tags and coloured fabric squares glued to their backs. Aerial videos of the calves held with their dams in a pasture were made over four days using UAVs. Recordings were made during two periods on most days of study: Morning-afternoon (M-A) and Evening (Eve) (Poor weather precluded data collection in 2 of the possible 8 periods.) Calves were acclimated to the UAVs by flying one UAV intermittently overhead at an approximate height of 20 m on the days before experimental recording began. We used two UAVs to video-capture the following: (1) the location of all individuals (UAV flown at 100 m) and (2) the identity of cow-calf pairs (UAV flown at 30 m). Still-images extracted from the UAV-acquired video screenshots were used to produce orthomosaics using Agisoft Photoscan software. An orthomosaic is a large composite image that has been processed to remove lens and perspective distortions in the original images. The orthomosaics captured all of the cows and calves in a single image, from which we measured cow-calf proximity with ImageJ software. Average cow-calf distances were then calculated for each of the six periods of data. Mean cow-calf proximity varied significantly among the six time periods (ANOVA; F=25.33, P<0.01), with M-A periods forming a group distinct from Eve; the average distance between cows and their caves was less in the M-A periods (`X=33.2, SD=5.74) than the Eve (`X=67.9, SD=7.59). These results will be compared with global positioning system (GPS) measurements on the same animals to examine the use of UAVs as an effective tool in animal welfare and behavioural studies in field conditions.

Differences in behaviour between aubrac and Lithuanian black and white bulls kept for beef

Vytautas Ribikauskas[1], Jūratė Kučinskienė[1], David Arney[2] and Daiva Ribikauskienė[3]
[1]Lithuanian University of Health Siences, Veterinary Academy, Tilžės 18, 47181 Kaunas, Lithuania, [2]Estonian University of Life Sciences, Institute of Veterinary Medicine and Animal Sciences, Kreutzwaldi 62, 51014 Tartu, Estonia, [3]Kaunas University of Applied Sciences, Pramones pr. 20, 50468 Kaunas, Lithuania; vytautas.ribikauskas@lsmuni.lt

In Lithuania, there is a tendency to have mixed cattle breed farms, where dairy farmers support the financial viability of their businesses with beef cattle as an additional production source. In such cases, all male offspring of dairy cows are kept for veal in the same conditions as hybrid or purebred beef bull calves. This situation raises some questions about the welfare and behavioural needs of such different animals. This study aimed at comparing the behaviour of finishing bulls from two different breeds – Lithuanian black and white (dairy cattle breed, LBW) and aubrac (beef cattle breed). Two groups of 11 to 14 month old bulls, one of the Aubrac (n=17) and the other group of the LBW breed (n=22) have been observed. At the start of the experiment, average weight of Aubrac and LBW bulls was 372.7±8.9 and 296.9±5.7 kg respectively. Animals were raised in insulated barn in pens with deep litter, both breeds under similar conditions (animals were 11 to 14 month old, each group in separate 100 m^2 pen). The behaviour was observed for a period of 24 hours (00:00-24:00 h) in June. The bulls were observed using instantaneous sampling with a 5-min sampling interval. The following behaviours were considered: total lying, eating feed at the feeding trough, drinking, standing ruminating, lying ruminating, walking, standing idling, lying resting and aggressive behaviour. In the present study, we found that total lying time of Aubrac and LBW bulls was 652 min (45.3±3.0%) and 677 min (47.0±3.0%) per day respectively, total standing time 720 min (50.0±2.8%) and 692 min (48.0±2.8%) per day respectively. The study indicated that there were no significant differences between the two groups of animals as regards moving behaviours. Aubrac bulls spend more time in standing idling (17.2±1.5 vs 12.5±1.1%, P<0.025), standing ruminating (10.7±1.1 vs 7.9±0.8%, P<0.05), drinking (1.6±0.3 vs 0.7±0.2%, P<0.01) and less time in eating (20.5±2.0 vs 27.1±2.3%, P<0.05) than the LBW bulls. LBW bulls spent more time lying with less lying bouts (6 lying bouts) compared with aubrac bulls (9). Aubrac bulls were distinguished by more clearly expressed aggressive behaviour in the group. The average time per animal spent in fighting accounted for 4.2% of the total time (39 encounters between two or more bulls) compared with the average 1.5% of time (21 encounter) for LBW bulls. At 90 and 95% thresholds, behaviour synchronization of Aubrac bulls was lower compared to LBW bulls. That means that 90 (and more) and 95% (and more) of Aubrac bulls were observed showing the same posture (lying or not lying) 39.6 and 25% of all observations accordingly compared to 48.6 and 38.9% (P<0.025) of observations of LBW bulls. These differences could be related to different breed of animals, and should inform decision making about the management of the two breeds studied.

Effect of switch trimming on hindquarters cleanliness of cows

Ken-ichi Takeda[1] and Mao Shibata[2]
[1]Shinshu University, Institute of Agriculture, Academic Assembly, 8304 Minamiminowa, 399-4598 Nagano, Japan, [2]Shinshu University, Faculty of Agriculture, 8304 Minamiminowa, 399-4598 Nagano, Japan; ktakeda@shinshu-u.ac.jp

Tail docking of cows is performed in some countries to improve cow cleanliness and udder health. Periodic trimming of the long hairs growing at the distal end of the tail (i.e. switch trimming) has been proposed as a humane alternative to tail docking in cows. Here we investigated the effect of switch trimming on hindquarter cleanliness of cows. Five Japanese Black cows, aged 9-14 years, were repeatedly used in three treatments. The length from the tail head to the switch tip was 105 cm in the control treatment. In addition, two experimental treatments were established: (1) a simulated tail docking treatment that assumed the length from the tail head to the switch tip to be 20 cm and the switch tip was painted; and (2) a switch trimming treatment in which the length from the tail head to the switch tip was approximately 60 cm. Cows were moved into the experimental pen sequentially. The same amount of paint was applied to the end of the tails in all treatments. In the simulated docking treatment, we painted the switch of each cow twice using a brush to ensure the same amount of applied paint as for the other treatments. After 5, 10 and 20 min of paint application, the side and rear side of each cow was photographed. We repainted the switch after 5 and 10 min. Each cow was washed with a special shampoo used for cattle shows before the experiment and after each treatment to remove the paint. During the treatment, to standardise the number of tail swings in each treatment (20-30 swings/min), we simulated flies landing on the back of each cow using toy flies attached to a mobile. In each treatment, the paint adhesion rate on the cow increased with increasing time. The paint adhesion rates were significantly different between treatments (P<0.05, F2, 12=14.5, one-way repeated ANOVA). Using a Tukey-Kramer test, a significantly higher paint adhesion rate was observed in the control (70.5±8.3%, P<0.05), whereas it was lower in the switch trimming treatment (34.8±3.3%, P<0.05) and lowest in the simulated docking treatment (2.4±0.7%, P<0.0.5). Our results show that the simulated tail docking treatment the most effective for transfer dirt to body, but considering that the function of the tail is lost in a docking, it can be said that the switch trimming is an effective way to guarantee the cleanliness and welfare of cows.

Prevalence of lameness and skin lesions on intensive dairies in southern Brazil

Joao H.C. Costa[1,2], Maria J. Hötzel[2], Tracy A. Burnett[1], Angélica Roslindo[2], Daniel M. Weary[1] and Marina A.G. Von Keyserlingk[1]

[1]*University of British Columbia, Animal Welfare Program, Vancouver, BC, V6T 1Z4, Canada,* [2]*Universidade Federal de Santa Catarina, LETA – Laboratório de Etologia Aplicada, Florianópolis, SC, 88034-001, Brazil; jhcardosocosta@gmail.com*

Lameness is one of the greatest welfare challenges facing the North American and European dairy industries but there is little information on the prevalence of lameness and lesions in the intensive Brazilian dairy industry. Our objective was to determine the prevalence of lameness and skin lesions on intensive dairies that used either freestalls, compost or a combination of these two systems. Data were collected in the autumn and winter months of 2016 from 50 dairy farms located in Paraná state, including 22 freestall dairies, 13 compost barn dairies and 15 dairies that used a combination (these dairies used freestalls for most of the herd but housed their vulnerable cows – transition and sick cows – on a compost bedded pack). Each visit to the farm consisted of three parts: (1) a management questionnaire, (2) inspection of the milking parlour, the barn where animals were housed, and the walkways from the barn to the milking parlor, and (3) an evaluation of all lactating cows as they exited the parlour for lameness, hygiene, BCS and skin injuries (lameness 1-5 score, hygiene 0-1 score, BCS 1-5 score, and skin lesions 0-1 score). The median one-way chi-squared test was used to compare production systems. We found no difference between farm types in foot health management practices or average daily milk production per cow (31 [29-33.9] kg/d; median [q1-q3]), percentage of Holstein cattle in the herd (100 [90-100]), conception rate (35.8 [30.2-38]%) or pregnancy rate (15 [13.7-18]%). The compost farms (85 [49.5-146.5] milking cows) were smaller than both the combined farms (360 [150-541.5] milking cows) and the freestall farms (270 [178-327.5] milking cows; P<0.01). The overall prevalence of severe lameness across all farms (score 4 and 5) was 21.2 [15.2-28.5]% but was lower on the compost farms (14.2 [8.45-15.5]%) in comparison to freestall farms (22.2 [16.8-26.7]%) and the combined farms (22.2 [17.4-32.8]%; P=0.03). Less than 1% of all cows scored on the compost farms were observed with swollen and/or wounded knees, which was lower than either the freestall or the combined dairies (7.4 [3.6-11.9]% and 6.4 [2.6-11.8]% of all cows scored, respectively; P<0.01). The same pattern was found for hock injuries, where the farm-level prevalence was 0.5 [0-0.9]%, 9.9 [0.8-15.3]%, and 5.7 [2.6-10.9]% on compost, freestall or combined dairies, respectively (P<0.01). No differences between farm type were observed for hygiene and BCS. On average, 2.7 [0.8-10.9]% of lactating cows had a soiled side, 15.4 [2.1-37.4]% had dirty legs and 1.7 [0-9.3]% had dirty udders. The average herd-level BCS across farms was 2.9 [2.9-3], with 0.52% of the cows having a BCS<2. Overall, the prevalence of lameness in intensive dairies in southern Brazil is high and highlight the need for changes in environmental design and management practices. We found that compost barn farms in this region had reduced lameness and injuries in relation to free stall and combined dairies.

Shelter use by horses and beef cattle during the winter

Katrine Kop Fogsgaard, Karen Thodberg and Janne Winther Christensen
Aarhus University, Department of Animal Science, Blichers Allé 20, 8830 Tjele, Denmark;
katrine.kopfogsgaard@anis.au.dk

In Denmark, horses and beef cattle can be pastured 24 h/day during the winter if they have access to shelters. For horses, specific requirements exist for individual space allowance but there are no regulations in terms of shelter design, e.g. the number of entrances and partitions inside the shelter. Similarly, for beef cattle, only general recommendations of space allowance and width of entrances to shelters exist. Currently for beef cattle, an ongoing project examines the effect of space availability per cow and the shelter design on time spent in the shelters. An initial study was conducted with 9 groups (3-7 individuals) of Angus cattle. Each group, all kept on pasture in one private herd, had access to a 50 m^2 shelter open on one long side with straw-bedding (5×10 m). Infrared cameras were placed inside all shelters taking a photo every 15 minutes 24 h/day. The first preliminary results from this study show that at least one individual was inside each shelter in 30-60% of a 14-day period (Dec-Jan). The shelters were mainly used in the dark hours. These results were based on a setting where space allowance per individual was larger than the Danish recommendations (4 m^2 per adult individual) which might have facilitated the use of the shelter. Based on this knowledge the following steps in the project is to investigate the effects of space allowance (4, 6, or 8 m^2 pr. adult individual), shelter design and weather conditions on shelter use by Angus beef cattle during the Danish winter. For horses, we investigated the effect of the number of entrances (1 vs 2), partitions inside the shelter, and use in relation to weather conditions. Thirty-two Icelandic horses participated. The horses were kept on pasture in eight groups of four, and each group had access to a shelter (30 m^2), which in the first study period (Dec-Jan) had either one or two entrances (n=4 groups in each treatment). In the second study period (Jan-Feb), all shelters had two entrances and half were equipped with partitions inside the shelter. Infrared cameras were used to assess shelter use, determined as % of pictures (6 pics/hour) where one or more horses was present inside the shelter. We found that groups with two entrances used the shelters significantly more than groups with only one entrance (Median Percentages [IQR]: 12.6% [7, 20] vs 3.0% [2, 4], Mann-Whitney U-test: P=0.029), whereas partitions inside the shelter did not affect shelter use. The shelters were used more when daily temperatures were below 0 °C compared to warmer days (P<0.001). Furthermore a positive correlation between shelter use and wind speed was found (r_s=0.5, P=0.02). Regardless of treatment, the shelters were mainly used at night (P<0.001). We conclude that shelters for horses should have at least two entrances.

Feeding behavior of group-housed calves in Midwest US farms with automated feeders

Marcia Endres, Mateus Peiter and Matthew Jorgensen
University of Minnesota, 1364 Eckles Ave, St. Paul, MN 55108, USA; miendres@umn.edu

It is becoming more common in the USA to house preweaned dairy calves in groups and feed them using computerized automated calf feeders. However, very limited research has been conducted in the USA to describe behaviors of calves when using these feeders and how they might be associated with housing and farm management factors. The objective of this observational, descriptive study was to characterize milk consumption, calf drinking speed, number of calf visits to the feeder (rewarded and unrewarded), and calf daily weight gain in 25 randomly selected farms in the Upper Midwest USA using automated feeders to feed their preweaned calves. Feeding behavior data were collected from the feeder software for approximately 18 months. We used PROC MEANS in SAS to calculate means and SD for each variable across all farms. Experimental unit for the analysis was calf-day (an average reading per calf/day recorded by the feeder software). We found that drinking speed (ml/min) was 794±324; n=54,747) with a mean/farm ranging from 442 to 1,113 ml/min. The average daily milk allowance (l/calf) was 8.7±2.3; calves consumed 87±21% (n=62,548) of their allowance resulting in an estimated daily milk intake of 7.6 l/calf. Mean estimated daily milk intake/farm ranged from 5.5 to 11.6 l/calf. The number of daily rewarded visits (visits when calf is entitled to receive milk; visits with milk) was 4.8±3.4 (n=53,798); mean/farm ranged from 2.5 to 6.9 visits; however, most farms averaged between 4 and 6 visits. The farm with greatest daily milk allowance in the study (15 l/calf) had a mean of 6.9 visits. The number of unrewarded visits (visits without milk) was 6.5±7.7 (n=53,798); mean/farm ranged from 1.0 to 9.9 visits. The farm with the greatest milk allowance had the lowest number of unrewarded visits (1/day). Number of unrewarded visits was greater than the number of rewarded visits (P<0.01) in this study. The number of unrewarded visits has been shown in previous studies to be associated with calf health along with daily consumption and drinking speed. Estimated daily weight gain (g/day) was 804±263 (n=60,205); mean/farm ranged from 567 to 1,131 g/day. The farm with the greatest milk allowance had estimated daily weight gain per calf of 1,131 g/day. These feeding behavior and weight gain measurements were most likely influenced by differences in housing and management practices across farms.

Effects of prenatal stress on lamb immune response, brain morphology and behaviour

Leonor Valente[1,2], Kenny M.D. Rutherford[2], Carl W. Stevenson[1], Jo Donbavand[2], Nadiah Md Yusof[2], Kevin D. Sinclair[1] and Cathy M. Dwyer[2]
[1]*University of Nottingham, School of Biosciences, Sutton Bonington, LE12 5RD, United Kingdom,* [2]*SRUC, Animal Behaviour and Welfare, SRUC, Roslin Institute Building, EH25 9RG, United Kingdom; leonor.valente@sruc.ac.uk*

Stressful conditions experienced in pregnancy can have a detrimental effect on progeny development. However, the extent to which common husbandry conditions experienced by housed pregnant sheep have similar negative impacts is not known. We hypothesised that conditions such as high stocking density and social instability during pregnancy would negatively affect lamb immune response, brain morphology and behaviour. At week 11 of gestation 77 twin-bearing Scottish Mule ewes were allocated, balanced for parity, to either Control (C, n=42) or Stress (S, n=35) groups of 7 ewes. Groups differed in lying area (C: 2.70 m²/ewe, S: 1.19 m²/ewe), feed space (C: 71 cm/ewe, S: 36 cm/ewe) and social stability (C: remained undisturbed, S: underwent two social mixing events). All ewes were placed under Control conditions one week prior to the start of lambing. At 7 weeks of age two sub-groups of lambs were selected, balanced for sex and dam parity, and submitted either to euthanasia for collection of Hippocampus(C=10, S=10) or to blood sampling for measurement of levels of Serum amyloid A (SAA) (C=20, S=20) before and after vaccination. At 13 weeks the lamb (C=7, S=7) behavioural response to five consecutive five minute isolations was assessed. Data were analysed with a linear mixed model in Genstat, using the Restricted Maximum Likelihood (REML) method. All experimental procedures were performed in a research farm setting and completed under the authority of the Home Office Animal (Scientific Procedures) Act (1986) with approval from the local ethics committee. Animal care was carried out according to UK legislation and Defra guidelines, and approved by the ethics committee. There were no differences in dendritic spine density between treatment groups on pyramidal neurons from hippocampus cornu ammonis (CA1) area. However there were two interactions between prenatal stress and lamb sex for stubby (F=7.6, P=0.016) and thin (F=8.0, P=0.014) spines in this brain structure. C males had a higher proportion of stubby spines than S males (mean ± se: C=0.42±0.017; S=0.36±0.015), whereas C females did not differ from S females (mean ± se: C=0.33±0.015; S=0.36±0.016). Moreover, thin spines showed a lower proportion in C males than in C females (mean ± se: M=0.38±0.019; F=0.51±0.017) but no differences were found in S males or females (mean ± se: M=0.43±0.017; F=0.44±0.018). S did not affect the SAA response to vaccination (F=0.31, P=0.579). S had no effect (P>0.05) on the number of vocalizations, activity or escaping attempts in the repeated isolation test. However, given the small sample size used behavioural data should be interpreted with caution. Maternal stocking density and social instability altered the neuronal structure of the lamb. Still, these changes are dependent on lamb sex. Importantly, they do not necessarily translate into detectable behaviour differences.

Effect of ammonia exposure on eating and ruminating behaviour in sheep during simulated long-distance transport by ship

Yu Zhang[1], Lauréline Guinnefollau[2] and Clive J.C. Phillips[1]
[1]Centre for Animal Welfare and Ethics, School of Veterinary Science, University of Queensland, Gatton, Queensland, 4343, Australia, [2]Université Francois Rabelais, Université Francois Rabelais, Tours, France; yu.zhang2014@gmail.com

Ammonia accumulation in sheep accommodation during live export is a major welfare problem. Ammonia is an irritant to mucosal surfaces, including the buccal cavity, which may be linked to previously observed reductions in feed intake. We investigated the mechanisms for this by studying eating and ruminating rates in sheep in a climatic control room that simulated the typical ammonia concentration. Twelve 6-month-old lambs (mean body weight 24.5 ± 2 kg) were randomly assigned to control or 30 ppm ammonia treatment (a validated maximum for livestock in live export) for three periods in a change-over design. Each period lasted for 14 days, which is a typical journey duration from Australia to the Middle East. Sheep were randomly allocated to individual pens. Animals were exposed to a 12 h light-12 h dark regime. Ammonia concentration, room temperature and humidity were measured three times each day. In each period, ruminating behaviour was recorded for ten periods of 60 seconds from 14:00 pm to 19:00 pm on day 7 and day 8. Rumination was recorded in chews/min and the time interval between swallowing and regurgitation. Sheep were fed 400 g TMR pellets and long-chopped lucerne hay *ad libitum* daily. A general linear model was used for data analysis in Minitab 16.0. Ammonia decreased hay intake (1.03 vs 1.10 kg/d, SEM=0.014, P=0.002). Ruminating chewing rate decreased with ammonia (85.68 vs 88.10 chewing bites/min, SEM=0.60, P=0.014) and the time between swallowing and regurgitation increased (6.75 vs 5.83 seconds, SEM=0.18, P=0.002). Palatability of a hard lucerne pellet and soft sorghum chaff was measured during a short (6 min) eating test. Ammonia reduced the rate of masticating lucerne pellets (2.20 vs 2.28 chews per second, SEM=0.012, P<0.001) but not sorghum chaff (2.26 vs 2.28 chews per second, SEM=0.018, P=0.47). Thus ammonia reduced feed intake and slowed chewing and swallowing, maybe due to an irritated buccal cavity. If proven, this would indicate reduced welfare of the very large numbers of sheep that are transported from Australia to the Middle East. This study was fully ethically reviewed to the standards required of ISAE.

Factors affecting misdirected dunging behaviour of loose housed lactating sows and their piglets

Janni Hales[1] and Vivi Aarestrup Moustsen[2]
[1]Jyden Animal Housing, Idomvej 2, 7570, Denmark, [2]SEGES Danish Pig Research Centre, Axeltorv 3, 1609 Copenhagen V, Denmark; jhp@jydenbur.dk

Commercial viability of farrowing pens for loose housed sows relies on good performance and proper functionality of pen. Pens must be designed to accommodate the sow and satisfy the behavioural needs of the sow. Partly solid floor provides a lying area and serves as nesting area that can maintain materials in the pen. When the sow has freedom to move around dunging behaviour is less controlled and soiled solid floors and poor hygiene can have a negative impact on the welfare of the animals as materials become dirty and risk of infection increase. Piglet dunging behavior is equally difficult to control and soiled floors can be caused by unwanted piglet dunging behaviour. Prevalence of unwanted dunging behavior is unknown and the aim of this study was to investigate how factors such as herd, season and time in lactation influence this behavior. Data was collected from two herds with 1,200 sows per year and Sow Welfare And Piglet protection (SWAP) farrowing pens, which is a farrowing pen with an option to confine the sows temporarily. Pens had 60% and 40% solid floor in herd 1 and herd 2, respectively. The floor was divided into three areas: 'creep', 'sow area' and 'under sloping wall'. Areas were scored for cleanliness ('clean', 'partly 'dirty' or 'soiled') and sow cleanliness was assessed ('clean', 'partly dirty' or 'dirty'). Cleanliness of floors and sows was analysed for effects of herd, season and lactation week using PROC GLIMMIX in SAS. Creep floor was clean in 93% of observations in herd 1 and in 95% of observations in herd 2 (P=0.04). Cleanliness of the sow area differed between herds (P<0.001) and across seasons (P<0.001). In herd 1 proportion of clean sow areas increased (P<0.001) from 63% in the week before farrowing to 89%, 82% and 85% in week 1, 2 and 3 after farrowing, respectively. Similarly, the proportion of clean sow areas in herd 2 increased (P<0.001) from 73% before farrowing to 92%, 86% and 86% in week 1, 2 and 3 after farrowing, respectively. Cleanliness of the floor under the sloping wall decreased through lactation in both herds (P<0.001), indicating that piglets developed unwanted dunging behavior. In herd 1 the area under the sloping wall was clean in 89% of observations before farrowing, decreasing to 72%, 54% and 41% in week 1, 2 and 3 after farrowing, respectively. In herd 2 this area was clean in 92% of observations before farrowing, decreasing to 71%, 37% and 37% in week 1, 2 and 3 after farrowing, respectively. Sows were cleaner in herd 2 compared to herd 1 (P<0.001) but for both herds cleanliness decreased through lactation (P<0.001). In herd 1 70% of sows were clean to lactation week 1, decreasing to 63% in week 2 and 55% in week 3. In herd 2 96% of sows were clean before farrowing, 94% were clean in week 1, 88% in week 2 and 87% in week 3. Results showed that herd 2 was generally cleaner than herd 1, possibly caused by different floor profiles. Results also suggest that unwanted dunging behavior of sows and piglets developed through lactation and indicated seasonal variation. Further research into factors that influence sow and piglet dunging behavior is warranted to improve hygiene in pens with solid floors in the latter part of lactation.

Selection for tameness affects metabolism and latency in a novel object test in junglefowl – a model for early domestication?

Rebecca Katajamaa, Beatrix Agnvall, Jordi Altimiras and Per Jensen
Linköping University, IFM Biology, Linköpings universitet, 581 83 Linköping, Sweden;
rebecca.katajamaa@liu.se

Domesticated animals are characterised by a coherent set of phenotypic traits. Tameness could be a central underlying factor driving that development. Therefore, we selected red junglefowl, ancestors of all domestic chickens, during five generations for high or low levels of fear of humans. Fear levels were determined through a standardised behaviour test, where the response to an approaching human was recorded. The birds were housed in an identical environment to ensure that the differences in fear levels between the two selection lines were mainly due to genetic effects. Previously, birds selected on only reduced fear of humans developed e.g. higher food competition and both larger eggs and offspring. So, in the fifth selected generation, we measured basal metabolic rate (BMR), feed efficiency and latency in a novel object (NO), to investigate if these traits had been affected by the selection on level of fear of humans. BMR was measured through open flow respirometry in 48 birds (six females and nine males from the low-fear line (L); 15 females and 18 males from the high-fear line(H)) at five to six weeks of age. The birds were placed individually in battery cages (60×40, height 40 cm) with *ad libitum* access to water and feed when 19 weeks old. Cages were furnished with dust bath, nest box and perch. Feed intake was measured for 7 days in 18 L-birds (8 female, 10 male) and 26 H-birds (14 female, 12 male). Average weekly growth of each bird between 112 and 200 days was divided by its total feed intake to be used as a proxy for feed conversion. At 21 weeks of age, the NO test was performed. The birds were presented with a cracked raw egg and when they started to eat from it, the NO was placed 10 cm from the egg. Latency to start feeding again was recorded with a maximum of 180 s. Generalized linear models were used for data analysis with selection, sex and their interaction as predictors. Wald X^2-statistics were used for determining significance. BMR was higher in L-birds (26.96 ± 0.29 ml O_2/min/kg in low-fear, 26.07 ± 0.26 ml O_2/min/kg in high-fear, $X^2=5.4$, P=0.02) with an effect of sex where females had a lower BMR ($X^2=17.1$, P<0.001). L-birds were also faster in the NO test (103 ± 17.4 s. for low-fear, 145 ± 12.5 s. for high-fear, $X^2=4.1$, P=0.044) with a sex effect where females were faster at approaching the novel object ($X^2=6.37$, P=0.012). There was a tendency towards higher feed efficiency in L-birds (growth/feed intake [%]; $5.0\pm0.5\%$ for low-fear, $4.1\pm0.3\%$ high-fear, $X^2=3.5$, P=0.061) but a significant interaction between selection and sex revealed that females mainly drove the effect ($X^2=10.5$, P=0.001). The results support the hypothesis that reduced fear of humans, a trait important for successful domestication, during the early phases of chicken domestication may have caused correlated selection responses leading to a more growth-promoting metabolism and faster animals in a NO test.

The ranging behaviour of laying hens in relation to their personality

Sabine Vögeli[1], Bernhard Völkl[2], Eimear Murphy[2], Luca Melotti[2], Jeremy D. Bailoo[2], Sabine G. Gebhardt-Henrich[1], Ariane Stratmann[1] and Michael J. Toscano[1]
[1]Center for Proper Housing: Poultry and Rabbits, Division of Animal Welfare, University of Bern, Burgerweg 22, 3052 Zollikofen, Switzerland, [2]Division of Animal Welfare, University of Bern, Länggasse 120, 3012 Bern, Switzerland; sabine.voegeli@vetsuisse.unibe.ch

Giving access to an outdoor range is considered a welfare-friendly way to keep poultry. However, previous studies on the individual ranging behaviour of laying hens showed that only a subset of animals use the range regularly, indicating the existence of sub-populations. The present study focused on individual ranging behaviour and investigated if differential ranging behaviour of laying hens is related to personality traits, specifically, fearfulness and risk taking. Four side-by-side pens, each containing 355 Brown Nick laying hens, were equipped with a radio frequency identification system, which allowed us to track the locations of 110 hens/pen: inside the pen or in one of three outdoor areas (veranda, bad-weather run, free-range). Based on the initial ranging behaviour of the birds, 15 hens/pen which rarely went outside (non-ranging, NR; n=60) and 15 hens/pen which often went outside (ranging, R; n=60) were chosen as focal birds for personality tests. The test took place on the verandas of adjoining pens in two experimental pens that served for acclimatisation and testing. Groups of 16 and 14 focal hens, respectively, were first placed in the acclimatisation pen, for at least 20 minutes and then individually tested in the test pen. During the first test (novelty-suppressed feeding test), a hen was placed in the middle of the test pen and the latency to approach a feeder in a corner and to eat a kernel of corn (max. 5 min; as a measure of fearfulness) was recorded. The test allowed for testing neophobia without previous training of the hens. After a second kernel was eaten, the hen was exposed to a falling object (startle test). The distance that the hen retreated from the feeder and the latency until the hen returned to the feeder and ate another kernel (max. 5 min; as a measure of risk taking) was recorded. Testing first occurred at 23 weeks of age, and then repeated at 30 and 42 weeks. Preliminary analysis of the novelty-induced feeding tests suggested that fewer NR hens went to the feeder compared to R hens, although this difference was not statistically significant (92 vs 115, P=0.12). In both groups, the number of animals that approached the feeder after being startled decreased from the first to the last test session (36 vs 31 vs 25 in NR and 45 vs 39 vs 31 in R), indicating learned avoidance to being startled. The latency to approach the feeder during the novelty-suppressed feeding test (P=0.86) and during the startle test did not differ between the two ranging groups (P=0.8) possibly due to a lack of food motivation rather than response to novelty. In summary, preliminary analyses do not suggest a relationship between behavioural responses and range use, at least with the respect to the conducted tests. Subsequent analyses comparing individual behaviour patterns between the test sessions, may lend further insight into whether and to what extent such a relationship exists.

The effect of dark brooders and overhangs on fearfulness and free-range use in slow-growing broiler chickens

Lisanne Stadig[1,2], T. Bas Rodenburg[3], Bart Ampe[2], Bert Reubens[2] and Frank A.M. Tuyttens[1,2]
[1]Ghent University, Salisburylaan 133, 9820 Merelbeke, Belgium, [2]Flanders research institute for agriculture, fisheries and food, Burg. v. Gansberghelaan 92, 9820 Merelbeke, Belgium, [3]Wageningen University & Research, De Elst 1, 6708 WD Wageningen, the Netherlands; lisanne.stadig@ilvo.vlaanderen.be

Broiler chickens with free-range access often make limited use of the outdoor area, perhaps because they are fearful to use the (entire) range area. There are indications that chicks reared with dark brooders, i.e. dark, sheltered heat sources that simulate a mother hen, may be less fearful later in life. Another possible reason why chickens are reluctant to go outside, relates to the transition from the house to the range, which may be too abrupt (e.g. a large increase in light intensity). This may be addressed by providing a more gradual transition by placing overhangs in front of the popholes. Therefore, we tested the effects of dark brooders (through a presumed long-term effect on fearfulness) and of overhangs on free-range use. The experiment was conducted using three production rounds with 400 mixed-sex slow-growing broiler chickens (Sasso T451) each. Birds were housed indoors in four groups of 100 from day 0-28. In each round, two groups had access to a dark brooder whereas conventional heating lamps were used in the two other groups. On days 22 and 23, an open field (OF) and a tonic immobility (TI) test were performed on 25 randomly selected birds per group. From day 28-70 the birds were housed in four mobile houses, in the same groups as indoors. These houses were positioned on an 80×80 m field, split into four plots. Each house had a free-range area that consisted of 50% short rotation coppice with willow (dense vegetation) and 50% grassland with artificial wooden shelters. Overhangs were placed in front of two houses and switched to the other two houses weekly (always one house with and one house without a dark brooder history). Free-range use was monitored three times daily (09:00, 13:00 and 17:00) on week days; the number of birds outdoors per shelter type was recorded, as well as their distance from the house (0-2, 2-5 or >5 m). On average, 32.4% of the birds were outside. Preliminary analysis showed no effect of dark brooders on the behaviour observed during the OF test, except for a trend on escape attempts (brooders: 24.4% vs no brooders: 38.6% of birds with ≥1 escape attempt; P=0.075). TI duration (brooders: 64 s vs no brooders: 76 s) and number of inductions (brooders: 1.6 vs no brooders: 1.5) were not affected by the presence of dark brooders. No effects of dark brooders or their interaction with shelter type, distance from the house, and age on free-range use were found. This may be due to a possible limited use of the dark brooders by the chicks. Alternatively, fearfulness may not be affected by rearing with dark brooders, although a relationship may still exist between fearfulness and free-range use. No effects of overhangs or their interaction with distance from the house was found on free-range use. Possibly their design was insufficient, or birds had no problems making the transition from inside to outside in the first place. Concluding, no significant effects of dark brooders and overhangs were found in this study.

Epigenetic marks of rearing conditions detected in red blood cells of adult hens

Fábio Pértille[1,2], Margrethe Brantsæter[3], Janicke Nordgreen[3], Luiz Lehmann Coutinho[2], Andrew M. Janczak[3], Per Jensen[1] and Carlos Guerrero-Bosagna[1]
[1]Linköping University, IFM Biology, Linköping University, 58 183, Sweden, [2]University of São Paulo, Animal Science, Av. Pádua Dias, 11, 9, Brazil, [3]Faculty of Veterinary Medicine, Department of Production Animal Clinical Science, Ullevålsveien, 72, Norway; fabiopertille@gmail.com

Stressful conditions are common in the environment where production animals are reared. Stress in animals is usually determined by the levels of stress-related hormones. A big challenge, however, is in determining past exposure to long-term stress of an organism. Epigenetics involves studying how environmental exposures affect gene regulation during the lifetime of organisms. It is expected that if animals are constantly subjected to stress and systemic hormonal changes, this exposure will imprint the epigenome of many cells types, including blood cells. Epigenomic effect of stress in blood cells have been reported in monkeys and humans. Chickens, however, provide a unique model to study stress effects in red blood cells (RBCs), which are nucleated in birds. Chickens are subjected to a variety of conditions in their production environment. The fact that chickens contain nucleated RBCs allows for a straightforward measuring of the epigenome in a cell type of easy access, and in live animals. The present study investigates in chickens whether two different rearing conditions can be identified by looking at DNA methylation patterns in their RBCs later in life. For this, RBCs were collected from 21 individuals of 24 weeks of age, reared in a system of open aviaries (AV) or in cages (CG). These rearing conditions are likely to differ regarding the stress experienced by the birds, as suggested by observations showing long-term differences in fearfulness and cognitive functions. DNA extraction from RBCs was performed using proteinase K. The sequencing library was prepared using the MeDIP method combined with an enzymatic genome digestion. The methylated DNA was sequenced using the Illumina platform. Data quality trimming was performed in paired-end short reads with the SeqyClean tool. The alignment of quality-trimmed reads was performed using the Bowtie2 tool against the chicken reference sequence (*Gallus_gallus* 4.0, NCBI). Coverage depth was checked using Samtools and a quality cut-off based on this parameter was stablished to avoid a skewed distribution of methylated sites along the genome. The Medips R-package was used for basic data processing, quality controls, normalization, and identification of differential coverage. Approximately 3.3 million 300 bp windows were analyzed between 4 AV and 6 CG individuals (which passed quality control). Of these, 61,911 regions (minimum coverage sum across individuals = 10) were tested for differential methylation (DMRs) between experimental groups. We found 115 DMRs in RBCs (P<0.0005). Network analyses of genes associated with these regions showed connections with important biological pathways using PathDB and Reactome tools, mainly related to immune system and signal transduction in opioid signaling. The aim of the present study is to generate a proof-of-concept for future detection of long-term stress in production animals using epigenetic measurements in cell types of easy accessibility in live animals.

Do laying hens have a preference for clean or unclean scratch mats?

Bishwo Pokharel, Luxan Jeyachanthiran, Ilka Boecker and Alexandra Harlander
University of Guelph, Department of Animal Biosciences, 50 Stone Road East, N1G 2W1, Guelph,
Ontario, Canada; pokhareb@uoguelph.ca

Enriched cages offer laying hens areas for scratching/foraging on a wire mesh floor. Scratch areas (mats) may cause an accumulation of droppings quickly, causing the mats to become unclean. Whether birds prefer clean over unclean mats has never been studied. The goal of this research was to determine the effect of scratch mat cleanliness on scratching/foraging behaviour of laying hens. A total of 288 laying hens was housed in 16 enriched cages (18 hens per cage), each cage with two identical compartments joined by a pop hole to allow free movement of hens between compartments. On a daily basis, half of the scratch mats (1 in each compartment, which accounted for ~14% of the total cage surface) was removed and cleaned, while the other half was cleaned and covered with 550 g of conspecific feces. Clean (C) and unclean (U) mats were then put back into the cages in a random order, avoiding side bias. Feed as litter substrate (~5 g per delivery per scratch mat per day) was delivered automatically onto the scratch mats by a spiral conveyor pipe. After identifying the time when hens were most active (mid-day), the number of visits and the total time spent scratching/foraging on C and U mats were video recorded for 10 min/day, 3 times a week, over a period of 4 weeks. The observation started when the feed was delivered on each single scratch mat. Data were analyzed using PROC GLIMMIX and presented as means±SE. There was a significantly higher ($P<0.01$) number of visits per minute to the U [3.5±0.12] than to the C [3.2±0.12] mats. Birds scratched/foraged on average for a longer ($P<0.05$) period of time on the U [106.1±4.95 sec] than on the C [96.7±3.51 sec] mats. These findings suggest that laying hens kept on a wire floor are motivated to forage on scratch mats covered in conspecific feces. Foraging on unclean scratch mats to get feed requires more effort to obtain the feed and might explain why laying hens kept in enriched cages showed a relative preference for foraging on unclean scratch mats.

Bovine bone is a valuable activity object in Finnraccoon (*Nyctereutes procyonoides ussuriensis***)**

Tarja Koistinen, Juhani Sepponen and Hannu T. Korhonen
Natural Resources Institute Finland, Halolantie 31 A, 71750 Maaninka, Finland;
tarja.koistinen@luke.fi

The knowledge of the interaction with activity objects is limited in Finnraccoon (*Nyctereutes procyonoides ussuriensis*). We measured the interaction with a biologically relevant activity object, namely bone, in juvenile Finnraccoons. The traditional cages (1.2 m^2) of sixteen sister pairs of Finnraccoon cubs were furnished with a bovine bone in July. The bones were removed from half of the cages (8 cages, Deprivation group) for two weeks in two occasions, in September and November/December. In both occasions, the interaction with the bone was video recorded, for 24 hours, before removing the bones and after returning the bones to the animals. The duration and frequency of oral and other kinds of interactions with the bone were registered by using continuous registered for the first 15 mins of each hour of the day (total of six hours of analysis in 24 hours). The data were analysed by using Linear Mixed Model (SAS). The Deprivation group interacted with the bones longer ($F_{1,14}$=8.12, P=0.0129) and more frequently ($F_{1,14}$=7.64, P=0.0152) than Finnraccoons having the bones available continuously. The deprivation treatment affected the interaction with the bone in the Deprivation group: the daily interaction increased from 491+-99 (mean+-SD) seconds in 42+-4 bouts to 1,732+-314 seconds in 90+-5 bouts in September, and from 270+-43 seconds in 35+-5 bouts to 1,645+-242 seconds in 55+-7 bouts in December (group-recording interactions: $F_{3,42}$=22.79, P<0.001, $F_{3,42}$=17.64, P<0.001), respectively. No such a difference between recordings (means: 418-715 seconds, 36-45 bouts) was observed in Finnraccoons having the bones continuously available. When extrapolated to a full 24-hour day, Finnraccoon pair interacted with the bone a total of 18-48 mins in 140-180 bouts in a day. In conclusion, Finnraccoons do not lose their interest in interacting with the bone. The deprivation treatment increases interaction, indicating a continuous need for an activity object. Bone is a valuable activity object in Finnraccoon.

The development and use of methods in animal welfare sciences: justifications, derivations, and concepts

Birgit Benzing and Ute Knierim
Kassel University, Farm Animal Behaviour and Husbandry, Nordbahnhofstraße 1a, 37213 Witzenhausen, Germany; benzing@uni-kassel.de

In animal welfare science a wide bunch of ethological and other indicators and methods to assess animal welfare has been developed. In order to explore the entanglement between ideal and practice of good science regarding the methodical development in animal welfare science, objectives were (1) to explicate the criteria for the choice of methods, (2) to identify those criteria which may legitimate a method as being established, and (3) to describe reasons for discrepancy between ideal and factual procedures. Structured guideline interviews were conducted with eight senior researchers, covering different European backgrounds and working in diverse areas of applied ethology and animal welfare science. They were asked about their concepts of animal welfare, the embedment of animal welfare within ethology, the development and choice of methods, as well as the relevance of methodological concerns for their own research. The interviews were carried out orally in English or German, recorded, transcribed and analysed with quantitative content analysis (software MAXQDA, VERBI GmbH). Code systems were developed inductively by two coders; disagreements were solved by consensus and the final code system checked for inter-coder-reliability. Results indicate that the choice of methods is clearly influenced by (all following categories were given by the interviewees) the scientist's 'background' and 'attitude', including welfare concepts, as well as 'research objectives', e.g. basic or applied research. Choices are also influenced by a 'random' factor, e.g. the work environment. Diverse views exist of when a method is regarded as established. While the validation of a method remains one major criterion, further criteria range from 'consensus' among the scientific community to 'being published' at least once in an adequate way. Discrepancies between ideal and factual procedures may result from 'insufficient resources' for proper validation studies and 'cost-benefit-considerations'. Unsatisfactory report and justification of the methodical choice was ascribed to 'common scientific sense', 'obvious knowledge' and 'the pressure to be keep papers short'. However, the interviewees explained inconsistencies also with 'sloppy science' and following scientific 'trends'. While some explanations by the interviewees apparently stand to reason, closer analysis revealed how subliminal considerations, too, may shape the development and handling of methods. For instance, the interviewees acknowledged the influence of underlying welfare concepts, whereas the role of methodological presumptions, such as scientific paradigms – e.g. giving more credit to physiological or quantitative indicators – remained a largely unrecognised issue. This will be further investigated in relation to the evolvement of the methodical spectrum.

Key-note:
Dogs, humans and genes: evidence for mutualistic symbiosis?

Per Jensen, Mia Persson and Ann-Sofie Sundman
Linköping University, IFM Biology, Linköping University, 58183 Linköping, Sweden;
perje@ifm.liu.se

During the course of evolution, certain species have come to form close inter-relationships with each other, favoring the fitness of both. Due to intentional and natural selection, individuals with genetic setups facilitating this relationship are positively selected. The dog is the oldest domesticated species and we suggest that the cohabitation of humans and dogs may be guided by similar genetic mechanisms in both species. Earlier research has demonstrated that dogs have evolved phenotypic traits differentiating them from their ancestors, the wolves, apparently in a direction facilitating life with humans. For example, they have a greatly improved ability to digest starch, and their ability to communicate with humans vastly exceeds that of wolves. The vocal repertoire of dogs is considerably larger than that of wolves, and most humans appear to have an innate understanding of their signals. To study the genetic underpinning of human-directed sociality, we measured the propensity of dogs to seek human contact and assistance when pursuing an unsolvable problem (trying to open a closed lid for a treat). After behavioural testing, DNA was obtained with buccal swabs. Using genome-wide association studies on 190 kennel-raised beagles, we found that their propensity to interact with humans for cooperation was significantly affected by two loci on canine chromosome 26 (GLM; $P<0.001$). Dogs homozygous for the A allele at one of the SNP-markers spent more time close to humans during the test than those homozygous for the C allele (GLM; $P<0.001$). The two loci identified are linked to five genes, four of which have previously been associated with social disorders in humans. In an independent replicate, we analysed the same genetic markers in two other breeds (golden and Labrador retrievers), and verified similar effects of the same loci (GLMM; $P<0.001$). These results suggest that social behaviour is affected by the same genes in both species. Furthermore, previous research shows that oxytocin may promote intra- as well as inter-specific social affiliation in both dogs and humans. We genotyped 60 golden retrievers on a previously identified SNP-marker associated with the oxytocin-receptor gene OXTR. The dogs were given oxytocin or saline control intra-nasally and then tested in the same problem-solving test as described above. Dogs with the AA-genotype at the marker increased the social contact with the owner after oxytocin treatment, while the opposite was found in dogs with the GG-genotype (GLMM, interaction treatment × genotype: $P=0.007$). The results show that human-directed social behaviour of dogs is affected by genetic pathways also involved in sociality in humans. This could suggest that mutualism may have evolved by selection of homologous genetic variation in dogs and humans.

Rat tickling: a systematic review of methods, outcomes, and moderators

Megan R. Lafollette[1], Marguerite E. O'Haire[2], Sylvie Cloutier[3] and Brianna N. Gaskill[1]
[1]Purdue University, Animal Sciences, 610 Purdue Mall, West Lafayette, Indiana, 47907, USA,
[2]Purdue University, Center for the Human Animal Bond, Comparative Pathobiology, 632
Harrison Street, West Lafayette, Indiana, 47907, USA, [3]Canadian Council on Animal Care,
Assessment and Accreditation Section, 190 O'Connor St, Suite 800, Ottawa, ON, K2P 2R3,
Canada; lafollet@purdue.edu

Rats initially fear humans, which can increase stress, decrease welfare and impact human-animal interactions. Rat tickling is a conspecific play technique that can improve human-animal interaction and rat welfare. However, experiments use a range of protocols assessing different outcomes. Without a synthesis of previous experiments, researchers have insufficient guidance for using tickling which may prevent its widespread adoption. Our objective was to provide a comprehensive overview of empirical tickling research to provide direction for future application and investigation. Our specific aims were to synthesize rat tickling protocols, outcomes, and factors moderating these outcomes. We consulted the Preferred Reporting Guidelines for Systematic Review and Meta-Analysis (PRISMA) and Systematic Review Centre for Laboratory Animal Experimentation (SYRCLE) guidelines. These provide evidence-based guidelines for defining inclusion/exclusion criteria, data extraction, and bias assessment. In our review, two independent investigators systematically evaluated all articles about tickling identified from PubMed, Scopus, Web of Science, and PsychInfo. Inclusion criteria were publication of original, empirical data on rats using tickling in a peer-reviewed journal. Bias was assessed using the SYRCLE bias tool. We identified 32 articles (56 experiments) about rat tickling published in peer-reviewed journals. The most common experimental method was to use the 'Panksepp Tickling Method' of 15 s baseline rest followed by 15 s of dorsal contacts and pins for a total of 2 min for at least 4 days (46%, n=26). Twenty-one experiments assessed tickling compared to a control. Although a variety of assessment measures were used, the most commonly found outcomes were increased positive vocalizations (n=16), increased approach behavior (n=9), improved ease of handling (n=5), decreased anxiety behavior (n=5), and decreased levels of stress hormones (n=2). Thirty-seven experiments assessed moderators of tickling outcomes; the most commonly identified moderators were individual differences (n=11) and housing type (n=8). Experiments minimized potential bias most commonly using blinding during outcome assessment (50%, n=28), having complete or addressing missing outcome data (41%, n=23) and ensuring similar baseline characteristics between groups (16%, n=9). We conclude that tickling is a promising method of human-animal interaction that can be used to improve rat welfare. Tickled rats usually experience increased positive affect and decreased anxiety and fear during general housing and handling. However, a wide variety of methods were used and techniques to mitigate bias were poorly reported. Establishing tickling best practices is essential since tickling outcomes are moderated by several factors. In particular, future research is needed to determine the effect of individual differences on rat welfare outcomes from tickling.

Cattle handling by untrained stockpeople gets worse along a vaccination work day

Maria Camila Ceballos[1], Aline C. Sant'anna[2] and Mateus J.R. Paranhos Da Costa[3]
[1]Grupo ETCO. Programa de Pós-Graduação em Zootecnia, FCAV-UNESP, Jaboticabal, SP, Via de Acesso Prof. Paulo Donato Castelane, S/N, 14884-900, Brazil, [2]Departamento de Zoologia, Instituto de Ciências Biológicas, UFJF. Grupo ETCO, Rua José Lourenço Kelmer, S/n, 36.036-330, Brazil, [3]Departamento de Zootecnia, Faculdade de Ciências Agrárias e Veterinárias, UNESP. Grupo ETCO, Via de Acesso Prof. Paulo Donato Castelane, S/N, 14884-900, Brazil; mceballos30@gmail.com

The aim of this study was to evaluate whether the quality of cattle handling changed as a consequence of the passage of working time during handling routines in trained and non-trained beef cattle farms. The quality of handling was assessed in 15 beef cattle farms at Pará state, Brazil by recording the behavior of 88 stockpeople (all males) when carrying on the vaccination procedure. The farms were classified according to the access to training as: (1) farms with systematic training programs (n=43 stockpeople from 9 farms) – defined as farms in which the stockpeople regularly attended trainings on good handling practices and animal welfare; (2) farms without any type of training (n=45 stockpeople from 6 farms) – defined as farms on which none of the stockpeople had received any training or prior information on the topic. The actions of each stockperson in each farm were assessed during one day of work (180.2±70 minutes, on average) by recording the positive (PA, defined by the sum of correct use of handling flag and frequencies of touch and positive vocalizations), and negative actions during handling (NA, sum of the frequencies of hitting the gate against the animal; hitting and prodding the animals with a wooden stick; negative vocalizations and twisting the animal's tail), then the total of PA and NA was summarized per farm, resulting in the variables FPA and FNA. Simple regression analyses were done to evaluate the effect of time of handling (each 15 min of work) on all stockpeople behaviors as response variables (FPA, FNA). The averages (± SD) of FPA were 145.5±82.3 and 72.9±56.8 actions per farm, for trained and non-trained farms, respectively, while FNA were 82.1±84.6 and 159.2±84.2 actions. At trained farms the stockpeople worked, on average, 180±70 min a day and 175±77 min at non-trained. For trained farms, no significant effect of time of handling was found (P>0.05, for FPA and FNA) while for the non-trained group, a significant effect was observed on FPA (P<0.05) and a trend on FNA (P=0.065). As 15 min of handling passes within a day, positive actions decreases (0.06±0.02) and negative actions increases on the order of 0.04±0.02. Our results suggest that cattle handling get worse over time when the stockpeople do not have any information about good cattle handling practices, what could be worsened by the fatigue of those people. In conclusion, this study highlights the importance of adequate workday hours associated with professional training and capacitation to minimize the poor quality of cattle handling in beef cattle farms, with beneficial consequences for animal welfare, labor safety and productivity.

The behavioural and physiological responses of horses to routine handling

Jenna Kiddie[1], Emma Grocutt[1], Sophia Kelly[1], Melissa Robertson[1] and Alison Northrop[2]
[1]Anglia Ruskin University, Life Sciences, East Road, Cambridge, CB1 1PT, United Kingdom,
[2]Nottingham Trent University, Brackenhurst Campus, Southwell, Nottingham, NG25 0QL, United Kingdom; jenna.kiddie@anglia.ac.uk

There is limited understanding of the more subtle behavioural responses demonstrated by the domestic horse during procedures such as tacking up and handling. It is possible that subtle behaviours are misinterpreted by humans and this may significantly affect horse welfare. The use of non-invasive methods to assess physiological responses could help to corroborate these behavioural responses. This pilot study aimed to investigate the response of horses to routine handling procedures by assessing subtle behaviours and physiological (right and left eye and ear temperature differences measured by infrared thermography (IRT)) indicators. The objectives of this project were to validate physiological and behavioural parameters that have the potential to indicate positive or negative responses by the horse. Ten horses were sampled from a local college's teaching yard. Horses were randomly assigned into two groups and were assessed using focal-animal sampling for the duration of two consecutive three-minute procedures using a cross-over design to reduce order effect: (1) horses were approached from their left and the routine handling of tacking up was carried out (assumed negative valence); (2) horses were approached from their left and allogrooming was carried out by the handler (assumed positive valence). Procedures were carried out in a tie-up area and recorded via four cameras, set up to capture the front, side and back of the horse, and an IRT camera in front of the horse (with an observer standing behind this camera). Wilcoxon rank-sum tests showed no significant treatment order effect. The following state behaviours varied significantly between valence treatments: mouth open (V=0, P=0.031), masticating (V=0, P=0.008), right ear forwards (V=3, P=0.005), head lowered (V=4, P=0.014), lean away (V=0, P=0.016), tail raise (V=0, P=0.004), and tail flick (V=0, P<0.001). The frequency of blinking also varied significantly between valence treatments (V=4, P=0.007). All of these behaviours were exhibited for longer or more frequently during the negative valence treatment. Temperature differences between left and right eyes and ears were also significantly different between valence treatment (eye: V=10, P=0.021; ear: V=10. P=0.021), with the right eye and ear being warmer than the left during the negative valence treatment. These results suggest that subtle behaviours could be used to differentiate between different emotional valences and lends support to the use of infrared thermography as a novel, non-invasive tool for measuring emotional response in animal welfare studies.

Tourism and marine mammal welfare

Donald Broom
University of Cambridge, Veterinary Medicine, Madingley Road, Cambridge CB3 0ES, United Kingdom; dmb16@cam.ac.uk

A review of the scientific literature shows that, for most marine mammals the demands and profits of tourism lead to reduced likelihood of killing, although there is a small demand for body parts. There are negative impacts on populations because animals are taken from the wild to attract visitors, including tourists, to zoos and aquaria and because of some live animal watching methods. The welfare of captive animals varies but is often poor. Performances for tourists can have positive or negative effects on welfare. Seeing captive animals may lead to more positive public attitudes to the animals and hence result in pressure to reduce both the poor welfare associated with hunting and killing in the wild and poor keeping conditions. Life expectancy is a welfare indicator and longevity of orcas in the wild is reported to be 60-70 years for females and 50-60 years for males. In captivity it is reported to be 25 years when infant mortality is taken into account. Median survivorship in the wild, according to one calculation, is 15 years and in captivity 14 years for captive born and 12 years for all individuals but there is some disagreement about these figures. Wild cetaceans, seals, sea-otters and sirenians are an attraction that leads to tourist boat trips. Boat approach for observation, chasing and swimming with the animals may cause avoidance behaviour, reduced time for foraging, social disruption, injury, increased disease or reduced breeding success: all indicators of poor welfare. However, well-controlled observation methods may cause no negative effects on the animals. Tourist expectations can result in more negative impacts but there is evidence for tourist acceptance and understanding of limitations on contact. Educational information has positive impacts on some tourist attitudes but other tourists do not want to receive it. Some quantitative information now exists on the extent of reduction in breeding success of seals and cetaceans, extent of avoidance of human swimmers, numbers of cetaceans with injuries likely to have been caused by boats and duration and extent of social group disruption. As a consequence, some aspects of codes of practice for tourist trips to see marine mammals can be founded on facts but others are a precaution related to general biological knowledge so are harder to enforce. Where biologically-based codes of practice are used, the net effect of tourism on animal welfare and populations can be positive.

Rethinking the application of ethological methods in animal-assisted-therapy research

Karen Thodberg[1], Bente Berget[2] and Lena Lidfors[3]
[1]Aarhus University, Department of Animal Science, Blichers Allé 20, 8830 Tjele, Denmark, [2]University of Agder, P.O. Box 422, 4604 Kristiansand, Norway, [3]Swedish University of Agricultural Sciences, Department of Animal Envoriment and Health, P.O. Box 234, 532 23 Skara, Sweden; karen.thodberg@anis.au.dk

Humans have always benefitted from the resources achieved, practical help and companionship of animals. Some studies even find general health effects of owning a pet, but not all, and some even find negative effects. The involvement of animals in a therapeutic setting, animal-assisted therapy (AAT) is getting more common, but even though the research area is growing, there is a lack of knowledge about causal pathways through which contact and interactions with animals can affect the mental state of mentally vulnerable persons. Among several suggestions for effective elements in AAT are physical contact with the animal; forming a relation to the animal; practicing skills of communication and gaining self-confidence through learning how to handle an animal. Traditionally the effects of animal assisted therapy are measured with validated psychiatric or psychological instrument such as self-evaluation or evaluation by a professional therapist as well as qualitative studies of clients with different mental and physical difficulties. In most cases the effects are measured as changes in psychiatric symptoms before and after a series of sessions. However, the relationship between such changes and the different element of the AAT is not identified this way. In this presentation we will explain how we think that future developments would benefit from cooperation between ethologist and researchers from psychiatry and psychology, and to a higher extent combine psychometric instruments with objective measures of the clients' immediate response using ethological methodology e.g. together with physiological measures. By using ethological methodology and physiological measures to quantify the human response in relation to the intensity, and type of human-animal interaction during AAT we are able to link the objective measures of immediate human response to the traditionally used measures of mental state. The use of ethology to study human behaviour is especially relevant when the target population has problems with verbal communication as is often the case in persons diagnosed with e.g. dementia or persons with autism. Much more focus on the response of the animals being part of different types of AAT is needed in order to be able to identify potential threats to the welfare of therapy animals. This approach calls for cooperation between several scientific disciplines and will help to move research in AAT forward and enhance the quality of future studies. The obvious advantage of cross disciplinary cooperation is the possibility to address scientific questions from several angles, ensuring a holistic approach, whereas a challenge could be to acknowledge and respect other research fields' use of different techniques and approaches, and to recognize that as a strength and not as an obstacle.

Behaviour of visiting dogs during a 6-week experimental trial – a pilot study
Anne Sandgrav Bak, Julie Cherono Schmidt Henriksen, Janne Winther Christensen and Karen Thodberg
Aarhus University, Department of Animal Science, Behaviour and stressbiology, Blichers Allé 20, 8830 Tjele, Denmark; annes.bak@anis.au.dk

The use of visiting dogs in nursing homes is widespread in the Western world. The visits affect the nursing home residents, but how these activities affect the dogs is not well studied. The aim of this explorative study was to study whether visiting dogs change their behaviour during a series of nursing home visits, depending on which resident they visit, and the behaviour of residents and handlers. The behaviour of four approved visiting dogs was observed during 10-minute visits involving a total of 28 residents in three nursing homes. Each resident was offered 12 visits during a 6-week period. The duration of lying, sitting and being active, and frequency of behaviour, which may be associated with frustration (yawning, scratching, licking lips, lifting paw) was registered. The duration of the residents' behaviour directed towards the dogs was recorded (physical contact, talking), as was frequency of corrections to the dog from the handler. Data were analysed by a Friedman two-way analysis of variance by ranks for effects of repeated visits (defined as t_1=visit 1-4, t_2=visit 5-8 and t_3=visit 9-12) and visits to different residents. Furthermore, the correlations between dog and resident/handler behaviour were analysed by a Spearman rank-order correlations, separately for t_1, t_2 and t_3. The statistical unit was a dyad, each consisting of one dog and one resident. Neither the percentage of time sitting, lying nor being active differed over time (P>0.05), however, prevalence of potential frustration behaviour decreased with time (χ^2 (df=2)=10.9; P<0.01), with the highest frequency per minute in t_1 (t_1=0.36 (range: 0-1.9), t_2=0.17 (range: 0-1.0), t_3=0.22 (range:0-1.4); t_1 vs t_2: P=0.001; t_1 vs t_3: P=0.07; t_2 vs t_3: P>0.05). In addition, the dogs changed their behaviour (frustration behaviour and postures) dependent on the resident being visited (P<0.001 for all behaviours). We found several significant correlations between the behaviour of the dog and the behaviour of the resident/handler, e.g. in t_2 and t_3 the duration of lying were negatively correlated to the number of corrections from the handler (t_2: r_s=-0.42, P=0.025; t_3: r_s=-0.51, P=0.008) and to the duration of talking and touching the dog by the resident (t_2: r_s=-0.41, P=0.028; t_3: r_s=-0.54, P=0.004). The frequency of potential frustration behaviour was positively correlated to the amount of corrections given by the handler in t_1 and t_3 (t_1: r_s=0.59, P=0.001; t_3: r_s=0.41, P=0.036). Overall, the dog's postures did not change over time, but was affected by the resident receiving the visit. This is in agreement with the correlations found between behaviour of the dog and behaviour of the residents. Behaviour related to frustration decreased over time and was positively correlated to the number of corrections given by the handler. It thus seems that the dogs are more frustrated in the beginning of the visit series compared to later on, and that situations, with less correction from the handler, are favourable.

What are we measuring? Aggression in shelter dogs violates assumptions of local independence and measurement invariance

Conor Goold and Ruth C. Newberry
Norwegian University of Life Sciences, Animal and Aquacultural Sciences, Arboretveien 8, 1432 Ås, Norway; ruth.newberry@nmbu.no

Studies of animal personality attempt to uncover underlying or 'latent' personality traits that explain broad patterns of behaviour, often by applying latent variable statistical models (e.g. factor analysis) to multivariate data sets. Interpreting behavioural responses as a function of the personality traits they are assumed to reflect, and making valid comparisons between individuals using personality traits, depends on two key, but infrequently confirmed, assumptions: (1) behavioural variables are independent (i.e. uncorrelated) after accounting for the effects of the latent personality traits being studied (local independence); and (2) personality traits are associated with behavioural variables in the same way across individuals or groups of individuals (measurement invariance). We investigated potential violations of these assumptions in observations of aggression in four age classes (4-10 months, 10 months - 3 years, 3-6 years, over 6 years) of male and female shelter dogs (n=4,743) in 11 different contexts. Data were gathered from a shelter's longitudinal behavioural assessment, and analysed as dichotomous variables according to whether or not dogs had shown aggression in each context whilst at the shelter. A structural equation model supported the hypothesis of two correlated (r=0.25; P<0.001) personality traits underlying aggression across contexts: aggressiveness towards people and aggressiveness towards dogs (comparative fit index: 0.97; Tucker-Lewis index: 0.96; root mean square error of approximation: 0.03). Intraclass correlations demonstrated that both traits were moderately repeatable (aggressiveness towards people = 0.479, 95% CI: 0.466, 0.491; aggressiveness towards dogs = 0.303, 95% CI: 0.291, 0.315). However, certain contexts related to aggressiveness towards people (but not dogs) shared significant residual correlations (P<0.05) unaccounted for by latent levels of aggressiveness. Furthermore, aggression towards people and dogs in different contexts showed interactions with dog sex and age, indicating that sex and age differences in aggression were not simple functions of underlying aggressiveness. In conclusion, our analyses have detected two sources of measurement bias in personality assessments, which could help to explain inconsistencies in the predictive validity of personality assessments in shelter dogs, especially regarding aggressiveness.

Which behavioural choices do dogs show during an animal assisted intervention?

Manon De Kort, Aimee Kommeren and Carlijn Van Erp
HAS University of applied sciences, Animal, Onderwijsboulevard 221, 5200 MA Den Bosch, the
Netherlands; kdm@has.nl

In the Netherlands the number of dogs performing as assistance dogs or in Animal Assisted Interventions (AAI) explosively increased in de the last decade. Research on AAI has shown many positive effects on humans, such as health benefits and psychological improvements. Science is still searching for the underlying mechanism. One step towards gaining more knowledge is to explore the process of the intervention. Little information is available on the perspective of humans and effects on humans. Information on the dogs' perspective seems to be lacking. Therefore this pilot study explored the process of an AAI aiming at the dogs' perspective. Which behavioural choices do dogs show during AAI? In this pilot study five dogs were filmed during their AAI at a nursing home for elderly with demential symptoms. These five dogs included 1 mixed-breed (M), 1 Labradoodle (M), 1 Stabyhound (F), 1 Kooikerhondje (F) and 1 Icelandic Sheepdog (F), all more than one year of age. All dogs were kept in family homes and commonly acted 1-2 a week as an AAI dog. The visits at the nursing home took place at 10.00 am (approximately 2 hours after feeding). The dogs arrived at 10.00 were they could relax for 20 minutes. After that the handlers brought the dogs to the clients. During the AAI the dogs stayed on a leash (1-1.5 m), on the floor, except when the client could not reach for the dog. Each dog visited a group of approximately 5 clients during 45 minutes. Once a week, for eight weeks, the AAI took place. Film footage of these visits in the first, third and last week were analysed using the Observer XT. The ethogram used, focused on behaviours in which certain choices of the dogs could be distinguished, such as: who initiates the state of being nearby or apart, who initiates contact, how do dogs make contact with the client and if the client and dog have contact how does the dog react? Since this was a pilot study and the number of dogs was limited only descriptive statistics were used. Two third of the time of the intervention the dogs were apart from the clients. Initiative for being nearby the client: by Handler 65%, by Dog 35%. Initiative for apart: by H29%, by D71%. When a dog is nearby a client, initiative for contact: H35%, D40%, Client25%. The dog prefers to contact the client by using its eyes or nose. When there was contact the dog could react with different behaviours, how he reacted, varied with who initiated the contact (Dog, Handler or Client). Some behaviours that were seen: no reaction (12% D; 34%H; 39%C), head turning towards client (24%D; 28%H; 25%C), body contact (6%D; 17.5%C) and sniffing (21%D; 21%H; 8%C). The different ways to get in contact with clients seem to be similar with the way dogs communicate with each other and humans according to other studies. The great number of distractions, such as other dogs, handlers, clients and food, did not seem to influence their preference of contact. Also the limited freedom of movement, due to the leash, could have influenced the results. However, when a dog was nearby a client he did choose to initiate contact. The dog most often chose to leave the session (apart). This was a pilot study with a limited sample size, more studies on this subject would be helpful to gain more knowledge.

Behavioural and physiological responses of two ovine breeds to human exposure

Mariana Almeida[1], Rita Payan-Carreira[1,2], Celso Santos[1] and Severiano R. Silva[1,2]
[1]*Animal and Veterinary Research Centre (CECAV), Universidade de Trás-os-Montes e Alto Douro (UTAD), Quinta de Prados, 5001-801 Vila Real, Portugal,* [2]*Departamento de Zootecnia, Universidade de Trás-os-Montes e Alto Douro (UTAD), Quinta de Prados, 5001-801 Vila Real, Portugal; mdantas@utad.pt*

The knowledge of temperament enhances management and animal selection, which translates to increased welfare and productivity. This study's intent was to verify the existence of differences in temperament between two sheep breeds – Churra da Terra Quente (CTQ) and Ile-de-France (IF) – by assessing the putative differences in behaviour applying an arena test with a human operator. After an initial one-week habituation period to housing and groups, ten adult female sheep, five CTQ and five IF, were tested during a three-week period. All animals were kept under the same conditions. For practical restrictions, only four animals per group were individually tested to ensure that individual behaviours were not being concealed by group behaviour. The arena was a 6×1.5 m open space, divided into equal parts of 1 m (zone zero to five) and the operator was static for ten minutes inside zone zero. Animal behaviour was filmed, and the footage was analysed while recording a total of 13 behaviours, including proximity and interactions with the operator. Blood samples were collected from 3 tested animal in each group, to screen cortisol serum levels (Immulite 1000, Siemens), immediately before and after the test (<2 minutes) by an experienced handler. These samples were collected into glass test tubes, centrifuged and plasma was transferred into Eppendorf tubes which were then stored at -70 °C. Data analysis consisted of factorial analysis and Fisher-PLSD test (Fisher's Protected Least Significant Difference) to compare means. Statistical analysis revealed significant differences ($P<0.05$) in locomotor activity, the IF being more active (2.86 vs 2.31 m/min). IF sheep also showed a higher number of loud bleats and a lower latency in bleating for the first time (48 vs 18 and 0.642 vs 4.9 minutes, respectively for IF and CTQ; $P<0.05$). Both groups displayed a reduction in locomotor activity ($P<0.05$) from week 1 to week 2 (3.76 vs 1.99 m/min) which might be explained by familiarisation with the operator. Results from cortisol analyses were consistent with existing studies, the IF sheep showing higher values (2.603 vs 2.096 μg/dl; $P<0.05$). Expectedly, a rise in cortisol levels was observed after the test (1.016 vs 3.684 μg/dl; $P<0.05$). No significant differences were found between breeds previously to test, but after the test the IF group presented higher cortisol values (4.128 vs 3.239 μg/dl; $P<0.05$), showing how the two breeds have different physiological reactions when exposed to humans. However, since the number of animals is reduced it would be necessary to conduct further experiments to determine differences between breeds, nevertheless the overall results seem to indicate that IF sheep are more reactive to the operator than CTQ, an autochthonous breed, demanding greater attention in its management.

Monitoring changes in heart rate and behavioral observations in donkeys during onotherapy sessions: a preliminary study

Emanuela Tropia, Daniela Alberghina, Maria Rizzo, Gemma Alesci and Michele Panzera
University of Messina, Veterinary Sciences, Polo Universitario dell'Annunziata, 98168, Italy;
emanuela.tropia@hotmail.it

Donkeys are used for assisted therapies (AT) in adults with mental health disorders where they serve as suitable and effective alternative to equestrian therapy. Studies on animal assisted therapy programs have shown that AT could benefit patients treated for schizophrenic disorders since animals could act as social facilitators which are able to increase motivation, mobility and interpersonal contact. Research on characteristics for selection of individual animals or to assess animal welfare during AT are limited. The aim of this study was to evaluate heart rate (HR) and behavioral characteristics of animals involved in donkey therapy (onotherapy) sessions. Six clinically healthy female donkeys (mean age 17.8 ± 5.7 years) were recruited in this study. Behavioral observations were made from 8 am to 8 pm for two consecutive days in order to compile an ethogram of diurnal activity. Subjects were monitored during two work sessions of 30 minutes with a control group (Group C) consisting of six healthy male adults without any previous riding experience and with a therapeutic group (Group T) of six male adults affected by schizophrenic disorders (code ICD-9 diagnosis code 295.8). During session work with both groups the same therapist modulated human-animal interaction. Each session was videotaped in order to score animal responses. Tactile and olfactory interactions between animal and patient, postural behaviors and intraspecific social interaction were evaluated and scored on a 5-point arbitrary scale (0 = very low to 4= very high). Heart rate was recorded during sessions using a Polar heart rate monitor (Polar Equine S-610i, Polar Electro Ltd., Warwick, UK) and transformed using the Polar Equine 4.0 Software. One of the donkeys scored poorly on the behavioral scale and was removed from the study and considered not suitable for AT (results from evaluation of behavioral observation were very low). Mann Whitney Test was used to analyse mean heart rate values and sum of the behavioral scores (Statistica 8 software). There were no statistical differences in behavioral indicators between groups, while mean HR during work sessions with Group T (70.40 ± 7.60) was significantly higher ($P<0.05$) than with Group C (58.60 ± 2.0). These preliminary results could be indicative of a different emotional state of donkeys probably due to a higher activity of patients (Group T) during riding. These preliminary results suggest that individual aptitude for AT in donkeys should be tested by validated methods and that HR monitoring, in particular it's variability (not considered in this preliminary study), could be useful to evaluate the emotional state of individual animals used in onotherapy. Measure of physiological indicators coupled with behavioral observation may help assess the animal's suitability reactions. Further research is necessary to assess the welfare of donkeys involved in AT.

Longitudinal stretching while riding (LSR) as a preventive training method of riding horses – will it improve welfare?

Linn Therese Olafsen[1], Inger Lise Andersen[1], Lars Roepstorff[2], Juan Carlos Rey[2] and Sylvia Burton[1]
[1]*University of Life Sciences, Animal-and Aquacultural sciences, P.O. Box 5003, 1432 Ås, Norway,*
[2]*SLU, Dep. of Anatomy, Physiology and Biochemistry, P.O. Box 7011, 750 05 Uppsala, Sweden;*
inger-lise.andersen@nmbu.no

The aim of the present preliminary project was to assess the immediate effects of a standardized method of longitudinal stretching of the top line with long reins during riding (LSR) on behavioral scores, mechanical nociceptive threshold (MNT; may quantify clinical neck and musculoskeletal sensistivity leading to dysfunction), pressure response score, and stride length, of 8 riding horses (aged 4 to 11, mixed breeds) subjected to a 20 minute treatment. MNT was measured at the 6 following locations, left and right of the mid line: muscle Brachiocephalicus at the basis of the neck, 1 and 2; mid thoracic Longissimus, 10 cm lateral of the dorsal midline, 3 and 4; between Tuber sacral and Tuber coxae on the mid Gluteus muscle, 5 and 6, with a pressure algometer. Behavioral expression of ears (from score 1: flat orientated backwards to 6: hanging in a relaxed state directed towards each side, eyes (from 1: eyes wide open showing much eye white to 6: no eye white, no tension around the eyes) mouth (from 1: large opening between upper and lower jaw, a lot of mouth activity with tongue or lips, rich in foam, lips very tensed, to 6: no mouth activity, closed, silent, relaxed mouth, very little or no foam), head-and neck position (from 1: head held high, tensed, nose above the vertical to 6: head and neck stretched downwards and forward with the mule below front knees, relaxed), collaboration with the rider (from 1: not willing to move forward, bucking, rearing, or kicking to 6: no negative behaviours towards the rider), working ability (from 1: low working ability to 6: high working ability), were scored by two pretrained observers. Four consecutive phases were evaluated: (1) first 5 minutes lunging; (2) first 10 minutes LSR; (3) last 10 minutes LSR; and (4) last 5 minutes lunging. Stride length was calculated using the ETB-Pegasus Digital Gait Analysis System. The last 10 minute LSR resulted in highest scores for all behaviours recorded compared to the other phases of the treatment, as for the total sum of scores (Genmod procedure: X^2-verdi=9.8, P-value<0.001). Horses with a high score for collaboration with the rider also had a lower, voluntary head position during riding (Polychoric corr. R=0.62; P<0.0001), a higher score for eye expression (Polychoric corr. R=0.43; P=0.016), a higher score for ears (Polychoric corr. R=0.57; P<0.0001), and a higher total sum of behavioural scores (Polychoric corr. R=0.82; P<0.0001), which may indicate a more positive mental state during riding. After the 20 minutes with LSR, MNT in five out of 6 locations along the top line were significantly different from before treatment (Genmod procedure: χ^2=588.1; P<0.0001). Pressure response score (0, 1 or 2) was significantly lower after LSR than before irrespective of location (Genmod procedure: χ^2=4.3; P=0.038). Stride length was significantly longer after LSR than before (Glm procedure: $F_{3,\,51}$=3.55; P=0.0208). In conclusion, these promising results suggest that we should further validate this method as a preventive riding method.

The influence of arena flooring type on behaviour in a test of calves' relationship towards humans

Stephanie Lürzel[1], Andreas Futschik[2] and Susanne Waiblinger[1]
[1]*Institute of Animal Husbandry and Animal Welfare, University of Veterinary Medicine, Vienna, Veterinärplatz 1, 1210 Vienna, Austria,* [2]*Department of Applied Statistics, JK University Linz, Altenberger Str. 69, 4040 Linz, Austria; stephanie.luerzel@vetmeduni.ac.at*

The behaviour of animals towards humans is often observed in a novel environment. Different degrees of novelty and resulting neophobia can be major confounders and the design of the arena may be essential for obtaining valid results. We investigated whether the presence of bedding has an influence on calf behaviour in a test of the animal-human relationship conducted in a novel arena. We re-evaluated behavioural data recorded from 82 calves that had or had not experienced 42 min of gentle interactions with humans during their first two weeks of life. The test arena measured 1.5×5.8 m^2 and had a concrete floor, including metal covers of the manure system. For half of the animals, the floor was covered with bedding (wood shavings) and not covered for the other half. The behaviour of the animals was recorded by two cameras placed above the arena. The test consisted of three phases of 3 min each: isolation, presence of an experimenter, isolation. They allow testing for a calming effect of human presence and for negative emotions induced by separation from the human. If the calf approached the experimenter, she stroked it for 30 s. For analysis, the arena was subdivided in five areas, weighted from 1 (at entrance) to 5 (farthest from entrance, in phase 2 with human). An area score was calculated as the sum of the products of the time (s) a calf was in a particular area with the weight of the respective area. The score ranged from 180 to 900, with lowest values indicating that the calf did not move far into the arena. There was a main effect of bedding on area score (mean±SD with bedding 527±166, without bedding 450±229; LMM, P<0.001) and frequency of entering area (10.9±8.3, 5.5±5.5; GLMM, P<0.001), tail flicking (7.6±13.1, 3.3±11.9; P=0.001), running/jumping (1.3±2.5, 0.3±1; P=0.006) and avoidance towards the experimenter (0.3, 0.7; P=0.04). Some of these effects were modified by a trend towards an interaction with phase or towards an interaction with treatment, e.g. frequency of running/jumping, both P≤0.08. For running/jumping, we subsequently ran a model for each bedding condition. No significant main effect of treatment but a trend towards an interaction of phase and treatment (P=0.06) existed in the model 'with bedding', with higher levels of running/jumping occurring in the treatment group than in the control group in phases 2 and 3. In contrast, there was a significant main effect of treatment (P=0.04) in the model 'without bedding', with more running/jumping occurring in the control group. The data show that the properties of the flooring in a novel test arena influence the behaviour of calves. The trends towards interactions with phase and treatment suggest that the results of an arena test may depend on the type of flooring used, and the example of running/jumping shows that the two flooring conditions can lead to partly opposite results regarding the effects of treatment.

Characterising personality traits and influences on these traits in German Shepherd Dogs

Juliane Friedrich[1], Pamela Weiner[1], Zita Polgár[1] and Marie J. Haskell[2]
[1]The Roslin Institute, University of Edinburgh, Easter Bush, Roslin EH25 9RG, United Kingdom,
[2]SRUC, West Mains Road, Edinburgh EH9 3JG, United Kingdom; marie.haskell@sruc.ac.uk

Understanding the expression of personality traits such as playfulness and fear in dogs, and what factors influence these traits, is important for the owners of pet and working dogs. In this study we assessed the effects of housing environment, level and type of training, the type of exercise regime and exposure to social and environmental stimuli on the expression of personality traits in German Shepherd Dogs (GSDs). Owners of GSDs were contacted by the UK Kennel Club and sent two questionnaires: a modified version of the C-BARQ canine personality questionnaire, consisting of 116 questions on behavioural characteristics such as aggression, anxiety, trainability and playfulness, and a 'management/lifestyle' questionnaire on care, exercise, training and housing of the dogs. There were 333 respondents to both questionnaires. Principal component analysis was used to derive personality traits from the C-BARQ responses. Linear models using backward selection of variables were used to assess relationships between dog characteristics, management factors and personality traits. The results suggested that 15 factors accounted for most of the variance in the behaviour questionnaire responses, including 'stranger-directed aggression', 'separation anxiety', 'stranger-directed fear' and 'playfulness'. Regression analysis showed that male dogs exhibited greater separation anxiety than females (t=-3.78, df=331, P<0.001), which has also been shown in Labrador Retrievers. Dogs trained to respond to a higher number of verbal commands had higher playfulness scores (t=6.97, df=330, P<0.001). This suggests that some aspects of the training experience may promote a positive emotional state in the dog, allowing playfulness to be expressed.

Social support from companion animals as an important factor influencing therapy and education effectiveness

Agnieszka Potocka

University of Warsaw, Faculty of Artes Liberales, Nowy Świat 69, 00-001 Warszawa, Poland; agnieszka.potocka@dogoterapia.net

This paper is a review of research and concerns about the role of companion animals working in Animal Assisted Therapy (AAT) and Animal Assisted Education (AAE). In this paper, based on a review of the research of social effects on humans of interaction with companion animals, I shall present the way to engage dogs, cats or other companion animals in AAT or AAE. Therapeutic interaction with animals could possibly increase effectiveness in therapy or education and without having negative impact on animal welfare. Students, patients or therapy clients who do not have sufficiently strong social support from family members, friends, and relatives may have weaker results in education in school or during any kind of their therapy than persons who have social support. Reason for this could be the feeling of isolation, ostracism, or lowered self-esteem. Dogs, cats and other companion animals can play similar roles as friends or family members and be an additional source of support. Cobb defined social support 'as information leading the subject to believe that he is cared for and loved, esteemed, and a member of a network of mutual obligations'. Social support is needed especially during crisis and difficult life situations, during education or therapy as a difficult relationship with therapist or teacher, fear associated with a new environment, new people, stress during exams or therapy procedures. Anthrozoology research strongly suggests that dogs' and other companion animals' company and trusting relationship with them could reduce anxiety, loneliness, stress and improve social contacts. The theoretical framework explaining the positive effect of human-animal relationship cover physiological mechanism of oxytocin release, behavioral system of attachment and social support. Recognition of significance of these mechanisms as components of therapy efficiency allows proper animal engagement. Satisfying social needs in relationships with others, gives benefits such as a sense of security, belongingness to a group, availability of help and, as a result, increased opportunities for survival, good health and a better quality of life. Dogs, cats, or horses should not be considered as merely tools in the therapeutic process, to be used in any way, without regard for their individual welfare. They should play the role of sensitive, valuable and sentient beings, with whom students, patients or therapy clients can establish a meaningful bond and receive social support from them. This approach facilitates better effectiveness of therapy or education and allows to maintain good welfare of animals.

The welfare of working dogs in cars

Lena M. Skånberg, Yezica Norling, Oskar Gauffin and Linda J. Keeling
Swedish University of Agricultural Sciences, Department of Animal Environment and Health,
Box 7068, 750 07 Uppsala, Sweden; lena.skanberg@slu.se

Working dogs, such as police dogs, spend a large amount of their time in cars. It has been hypothesized that a smaller cage could be safer for the dog during driving, especially at high speed, but that a larger cage could allow better comfort (e.g. for resting) when the car is parked. This presents a conflict. Our study investigated how dogs in a car are affected by cage size. Working dogs of the breeds German Shephard (n=8) and Springer Spaniel (n=8) from the Swedish Police and Customs were tested in four different cage sizes in four different car treatments; parked, slow cruising, normal driving and fast emergency driving. The study was conducted at a test track. Each dog was tested in all combinations of cage and driving style over a two-day period. Each combination involved 30 minutes in the car, except for the fastest driving style that was shortened to 5 minutes. The behaviour of the dogs was recorded by two video cameras in the car and the dog's heart rate and variability were recorded using a monitor (Polar V800). Statistical analyses used a mixed model or a Friedman test followed by a Wilcoxon Signed Rank test. Driving style had the largest effect on the dogs. The faster the driving, the higher the heart rate ($P<0.001$) and the lower the RMSSD values ($P<0.001$). Faster driving also resulted in a higher prevalence ($P<0.01$) of behavioural stress indicators (e.g. lip-licking, panting, shaking). No cage size seemed to be better regarding safety, such as decreasing the prevalence of slipping, losing balance or bumping against the walls of the cage. These behaviours occurred frequently, around 50% of the time, in the fastest driving style. Regarding comfort, the cage that was shorter than the length of the dog (although not the smallest in area), resulted in least time lying down during parking and slow driving ($P<0.05$) (on average 59% of the time) compared to the other cage sizes (on average 78% of the time lying down). Dogs also had a tendency to stretch more (3.75 times more), after spending 30 minutes in this cage ($P=0.058$). The largest cage, proportional to the size of the dogs, resulted in 2.6 times more changes in body postures ($P<0.05$). In conclusion, faster driving, including turns and sudden brakes, appears to be physically demanding and experienced negatively by dogs, irrespective of cage size. However, cage dimension still seems to matter regarding comfort. Transporting a dog in a cage shorter than its length, as measured from nose tip to point of buttock (currently allowed in the Swedish legislation for working dogs) could, due to worse comfort, negatively affect the performance of the dog when taken out of the car.

Do cull dairy cows become lame from being transported?

Kirstin Dahl-Pedersen[1], Mette S. Herskin[1], Hans Houe[2] and Peter Thorup Thomsen[1]
[1]Aarhus University, Department of Animal Science, Blichers Allé 20, 8830 Tjele, Denmark,
[2]University of Copenhagen, Department of Large Animal Sciences, Grønnegårdsvej 8, 1870
Frederiksberg C, Denmark; kirstin.dahlpedersen@anis.au.dk

Approximately 200,000 cull dairy cows are transported to slaughter in Denmark each year. Dairy cows are culled for a number of reasons and may be more vulnerable to transport than younger livestock. However, rules and recommendations on animal transport do not mention cull cows specifically and very limited research has focused on this particular group of animals. The aim of this study was to investigate whether transport of cull dairy cows by road for up to 8 hours affect their lameness score. The observational study involved 411 cull dairy cows, transported to slaughter during a 12 month period. The study included one slaughterhouse, trucks from four hauliers and cows from 20 dairy herds. Cows to be included were selected by the owners. An ethical permit allowed inclusion of varying clinical conditions, except for cows defined unfit for transport according to the EU regulation on animal transport (EC 1/2005). All management followed the normal, commercial procedures. The cows were examined clinically on-farm before loading, and then transported to Danish Crown's slaughterhouse in Holsted, Denmark. At the slaughterhouse, the cows went through a second clinical examination. Data were analysed in SAS and the McNemar test was used to compare the proportion of lame cows before and after transport. Logistic regression was used to evaluate the association between risk factors e.g. distance, body condition score and days after calving and the aggravation of locomotion score. Data were collected from 49 journeys. The mean distance was 129 km (range 20-339) and mean duration was 187 min (range 32-510). The mean number of pauses per journey was 2.0 (range 0-6) with a mean total duration of 48 min (range 0-155). The cows were culled after a mean of 2.9 lactations (range 1-10) and a mean of 270 days after calving (range 15-871). The majority of the cows (63%) were in normal body condition, 16% were thin (body condition score$\leq$2.5) and 21% were fat (body condition score $\geq$4.0). Only 9 cows arrived at the slaughterhouse in a clinical condition, which would have been considered unfit for transport according to current Danish practice. These were all severely lame (locomotion score 5). Before transport the prevalence of locomotion score $\geq$3 (5-point scale) was 31% and this was increased to 41% after arrival at the slaughterhouse (P<0.0001, df=1). Cows that were lame on-farm (locomotion score $\geq$3), had an incidence risk of 27% for becoming more lame, whereas cows that were not lame on-farm had an incidence risk of 16% for becoming lame. The maximum change in locomotion score was 3 points, but for 66% of the cows becoming lame or more lame during transport, the locomotion score increased with only one point. For 3% of the cows the locomotion score decreased with one point. Results show that cull cows can become lame during short road transport to slaughter. Further studies are needed to define thresholds for the risk factors and to examine possibilities for improvement e.g. management procedures in relation to culling and transports with special provisions in order to solve this challenge to animal welfare.

The effect of overstocking different resources within a freestall pen on the behaviour of lactating Holstein cows

Clay Kesterson, Randi Black, Nicole Eberhart, Erika Edwards and Peter Krawczel
The University of Tennessee, Knoxville, Animal Science, 2506 River Drive, Knoxville, TN 37996,
USA; ckesters@vols.utk.edu

Decreasing the availability of resting and/or feeding space has detrimental effects on lying behaviour. However, the interaction of overstocking pressure to varying degrees at these resources has yet to be evaluated to determine how cows may alter behaviour based on priority of use. The objective was to determine the effect of overstocking the freestalls, the headlock, or both at 160% on lying and social behaviour. Holstein cows (n=32) were housed in a naturally ventilated, freestall barn, fed a TMR at 07:00 and 15:30 h, and milked at 08:30 and 18:30 h daily. Groups (n=4) were balanced by parity (1.8±1.1) and DIM (224.3±58.5 d). Treatments were: 100% stocking density at freestalls (FS) and headlocks (HL), 100% stocking density at FS and 160% at HL, 160% stocking density at FS and 100% at HL, and 160% stocking density at FS and HL. Treatments were assigned using a 4×4 Latin square design with 14 d periods followed by a 3-d washout. Accelerometers were affixed to a rear leg of each cow to collect lying behaviours. Displacements 2 hours after each feeding were collected via video data. A mixed model (SAS 9.4) was used to evaluate the effects of treatment on total, left, and right side lying time, bout duration, and number of bouts, and aggression at the headlocks. Relative to 100% at FS and HL, overstocking the FS or overstocking FS and HL at 160% decreased lying time (13.2±0.4 vs 11.4±0.4 or 11.7±0.4 h/d, P≤0.01). Decreased lying time was also evident when FS or both FS and HL were overstocked compared to only overstocking the HL (11.4±0.4 or 11.7 vs 0.4 vs 12.7±0.4 h/d, P≤0.04). No differences in lying time was evident between stocking densities of 100% FS and HL, and 100% FS and 160% HL (P=0.19), or in the two treatments which included overstocking at the freestalls (P=0.47). No other lying behaviours differed (P≥0.15). Surprisingly, no differences in social aggression at the feed bunk were evident (16.6±4.8 n/d at FS stocked at 100% and HL at 160%; 12.8 n/d at FS stocked at 100% and HL at 100%; 8.3 n/d at FS stocked at 160% and HL at 100%; 6.9±4.8 n/d at FS stocked at 160% and HL at 160%; P=0.22). Cows were able to maintain lying time when overstocked at the HL alone, but lost lying time when overstocking was imposed at the freestalls. The lack of treatment effects on social aggression may have been driven by a greater priority given to accessing resting resources or from altering feeding patterns. This indicates that cows may be able to modify their feeding behaviour, but not their lying behaviour to mitigate reduced access to the associated resources. Management of freestall facilities may need to focus on maximizing access to resting space. However, cows enrolled on this study were more than 200 d into lactation, which may have reduced the importance of access to feed. Future studies should evaluate if prioritization of access to resting and feeding resources is shared by different stages of lactation.

Effect of water temperature on drinking behavior and stress response of sheep

Shin-ichiro Ogura, Iori Kajiwara and Shusuke Sato
Tohoku University, Graduate School of Agricultural Science, 232-3, Yomogita, Naruko-onsen, Osaki, 989-6711, Japan; shin-ichiro.ogura.e1@tohoku.ac.jp

Appropriate water supply is essential to certify animal welfare of livestock. Water temperature can affect the amount of water intake of animals; for example, cold water in winter may give stress to animals, and in contrast in summer, it may alleviate heat stress of them. Therefore in this study, the effect of water temperature on drinking behavior and stress response was examined under indoor feeding. The experiment was conducted on December (Cool season) and September (Warm season). The mean daily air temperature of each season was 2.6 ± 1.9 °C and 20.2 ± 2.7 °C, respectively. Eight lambs (mean live weight was 38.1-55.8 kg) were fed individually in *ad libitum* of alfalfa hay cube and water. Six animals among the eight ones were used for both seasons. In Cool season, two experiment was done; i.e. water temperature during habituation (5 days) was 25 °C (trial 1) and 5 °C (trial 2). In Warm season, the temperature during habituation (5 days) was 18-23 °C (trial 3). In the experiment period (4 days), water with different temperature (7, 15, 25 and 35 °C) was offered 09:30-13:30, then changed to the same temperature as in the habituation period. The treatment of each animal was changed every day by Latin square design. During the experiment period, water intake and dry matter intake was measured, and drinking behavior (number of drinking and its duration) was recorded by continuous observation. In addition in Warm season (trial 3), heart rate and cortisol concentration in saliva was measured every day in the experiment period, when the treatment of water temperature started. Actual water temperature offered to the animals in the experiment period was 5.5 (5.4-5.6), 18.8 (17.4-19.0), 26.0 (25.8-26.2) and 31.3 (30.9-31.9) °C in Cool season and 7.8 (7.0-8.5), 15.1 (14.7-15.5), 26.1 (23.0-29.1) and 33.5 (29.3-40.7) °C in Warm season. Water intake per live weight was 35 °C > 7 °C (ANOVA, $P<0.05$) in Cool season (trial 1 and 2), but there was no significant difference in water temperature in Warm season (trial 3). In both seasons, there was positive relationship between water intake and dry matter intake (Pearson's correlation analysis, Cool season: $r=0.472$, $P<0.001$; Warm season: $r=0.648$, $P<0.001$). In the treatment of 7 °C, water intake per drinking behavior was smallest, thus drinking rate was lowest (ANOVA, $P<0.01$). Although heart rate was not affected by water temperature, cortisol concentration in saliva was significantly high in 7 °C (1.12 ng/ml) than other temperatures (0.58-0.63 ng/ml) (ANOVA, $P<0.05$). These results showed that high water temperature (35 °C) increase water intake of animals in cool season, indicating that it increase dry matter intake. This study also suggested that water with low temperature (7 °C) gives stress to animals even in warm seasons.

Effects of genetic background and selection for egg production on food motivation in laying hens

Anissa Dudde[1], Lars Schrader[1], Mohammad Bashawat[2] and E. Tobias Krause[1]
[1]Friedrich-Loeffler-Institute, Institute of Animal Welfare and Animal Husbandry, Dörnbergstr. 25/27, 29223 Celle, Germany, [2]Leibniz Institute for Zoo and Wildlife Research, Alfred-Kowalke-Straße 17, 10315 Berlin, Germany; anissa.dudde@fli.de

The motivation to obtain food is a fundamental motivational state. Previous studies have shown that food motivation in chickens increases with food deprivation. However, it is also possible, that this motivation was modified during selection for egg laying productivity. We aimed to find out whether and to which extend hens of different layer strains have increased food motivation with respect to their annual egg productivity and whether they are eager to walk longer distances to reach food. We expect higher egg production to be linked with higher food motivation and, thus, faster locomotion over distances. We used 72 hens from a 2×2 factorial design, (2 genetic backgrounds, indicated by egg colour (white/brown) and 2 productivity level (low 200 vs high 300 eggs/y), thus incorporating four different strains. In their home pens, all hens had *ad libitum* access to food and water and, in addition, free access to grains presented in coloured bowls, which were used as food reward in the tests. For the test, hens were placed individually in a start box, leading to a corridor at which end the food reward was provided at increasing distances (50, 100, 200, 400 cm) on four subsequent days. At each distance, we measured the latency to reach the food reward and the velocity (cm/s). Data were analysed using Linear Mixed Effect Models (LME), with the factors body mass, genetic background, productivity, distances in the test (50, 100, 200, 400 cm), and their two-way interactions. Individual ID was a random factor. The latency to eat was significantly affected by productivity level, genetic background and distance (LME: productivity: $F_{1,65}$=6.59, P=0.01; genetic background $F_{1,67}$=8.84, P=0.004; distance: $F_{1,207}$=8.48, P=0.004; body mass: $F_{1,65}$=0.32, P=0.57). The velocity of hens differed significantly between the productivity levels and the genetic background (LME: productivity: $F_{1,65}$=6.08, P=0.01; genetic background $F_{1,65}$=9.06, P=0.004; distance: $F_{1,207}$=102.88, P<0.0001; body mass: $F_{1,65}$=0.45, P=0.5). All other not mentioned factors were not significant. Body weight was measured at the first test day and differed between white and brown layers (mean ± SD, brown: 1,878±204 g vs white: 1,634±179 g). High productive hens needed less time to feed and had a higher velocity compared to moderate productive hens, even when distance to food reward increased. These differences suggest a higher food motivation in high productive hens resulting from intense selection for egg productivity and their higher nutritional demand, rather than differences in locomotor abilities. Brown layers were faster than white layers. Although body mass was higher in brown layers, it did not affect velocity. Thus, differences in basal metabolism or in walking ability due to higher weights are unlikely be responsible for the higher velocity in brown layers. Instead, they can be explained by phylogenetic difference, which also has been demonstrated for other traits.

On-farm techniques used to minimize aggression in pigs by US pork producers

Sarah H. Ison, Ronald O. Bates, Juan P. Steibel, Catherine W. Ernst and Janice M. Siegford
Michigan State University, Animal Science, Room 1290, 474 S. Shaw Lane, East Lansing, Michigan,
48824, USA; shison@msu.edu

Despite decades of research, pig aggression remains a welfare issue. Although various management techniques have proved useful in minimizing aggression; it is not known to what level these techniques are implemented on-farm. One objective of an online survey to US pork producers was to find out what techniques are used, how useful they are perceived to be, and what are the potential barriers to their use. Respondents were given a list of techniques, were asked to select which ones they have used, and indicate how useful they perceived them to be (from 'very useful' to 'not useful at all'). They were also asked about the housing and management of pigs, and demographic information. Chi-square tests of homogeneity compared the frequencies with which different techniques were perceived as useful, and a step-wise binomial logistic regression analyzed the association between other factors and technique implementation (yes or no). Respondents (n=298), had nursery (11,490±4,013 pigs, range: 0-950,000), and/or finisher stage pigs (18,929±5,779, range: 0-1,500,000), and of these, some had both nursery and finisher pigs (n=181), and some had just nursery (n=24), or finisher (n=89) pigs. When asked if they regularly mix or re-group unfamiliar pigs, 36% of all respondents said yes (33% at nursery, 25% at finisher). Of all the respondents, 53% used techniques to minimize mixing aggression. These included pre-weaning piglet socialization (n=19), which was perceived to be more useful than all other techniques (P<0.05). Evening mixing (n=50), was perceived as more useful than odor masking agents (P=0.006, n=38), pre-exposure in a neighboring pen (P=0.03, n=45), and enrichment devices (P=0.008, n=41), but equally as useful as mixing all pigs to a new pen (n=108), barriers (n=25), or using specific mixing pens (n=21). Mixing pigs to a new pen was considered more useful than enrichment (P=0.04). Using mixed weight pens (n=9) was infrequent, but considered equally as useful as other techniques (P>0.05) by those who used it. The best regression model for predicting the odds of using a technique included the factors pig stage (nursery, finisher, or both), regular mixing (none, finisher, nursery, both stages) as significant predictors, and substrate/bedding use (yes, no, sometimes) as a tendency. Respondents with farms specializing in either nursery (log odds=4.08, P<0.001) or finisher pigs (log odds=4.42, P<0.001) only, increased the log odds of using a technique, whereas those that did not mix regularly at any stage (log odds=-5.03, P<0.001), reduced the odds. Those who selected yes to using bedding (log odds=0.88, P=0.077) tended to increase the odds of implementing techniques. These results indicate that, overall, producers avoid mixing. However, when mixing is required, techniques are used to minimize aggression, with pre-weaning socialization indicated as most useful.

If you give the pig a choice: suckling piglets eat more from a diverse diet

Anouschka Middelkoop[1], Raka Choudhury[2], Walter J.J. Gerrits[3], Bas Kemp[1], Michiel Kleerebezem[2] and J. Elizabeth Bolhuis[1]
[1]*Wageningen University, Adaptation Physiology Group, De Elst 1, 6708 WD Wageningen, the Netherlands,* [2]*Wageningen University, Host-Microbe Interactomics Group, De Elst 1, 6708 WD Wageningen, the Netherlands,* [3]*Wageningen University, Animal Nutrition Group, De Elst 1, 6708 WD Wageningen, the Netherlands; anouschka.middelkoop@wur.nl*

Stimulating solid feed intake in suckling pigs is important to ensure a successful weaning transition, exemplified by the correlation between pre- and post-weaning feed intake. In nature, piglets begin to sample food items in a playful manner when only a few days old. In pig husbandry, contrarily, suckling pigs are not encouraged to forage, whereas we hypothesize that this is crucial to increase piglet robustness via improved intestinal and gut microbiota development. One approach to familiarize pigs with solid feed at an early age might be by providing feed in a variety of forms, using diversity or novelty to stimulate the pigs' foraging behaviour. We studied the effect of dietary diversity (i.e. offering two diverse feeds simultaneously) vs novelty (i.e. regularly changing the flavour of one feed) on the foraging behaviour and feed intake of suckling pigs. We also hypothesized that piglets, rather than sampling from one feed only, would prefer a diverse diet. Piglets received *ad libitum* feed from 2 days of age in two feeders per pen. In treatment 1 (T1, n=10 pens) pigs were given feed A and feed B which differed in size, flavour, composition, smell, texture and colour. In treatment 2 (T2, n=9) pigs received feed A plus feed A to which additional novel flavours (4 different ones) were added from day 6 in a daily sequential order. Feeding behaviour was studied by weighing feed remains (d6, 12, 16, 22) and by live observations (4-min scan sampling, 6 h/d; d10, 15, 22; n=6 pens per treatment). Observations were also used to determine 'eaters' (i.e. pigs scored eating at least once). Data were analyzed using mixed models. Piglets did not prefer feed A (d2-22: 196±16 g/pig) or B (152±13) within T1 and did not have an overall preference for feed A with (d6-22: 78±4 g/pig) or without flavour novelty (66±6) within T2. In accordance, just a few piglets (T1: 1.5% and T2: 3.2% out of all eaters per treatment) were observed eating only one of the feeds throughout lactation. Interestingly, T1-pigs (d2-22: 327±28 g/pig) ate much more than T2-pigs (147±9; P<0.0001) and explored the (feed in the) feeders 2.6 times more at d15 (P=0.001). This also implies that feed A, the common feed provided in T1 and T2, was more consumed in T1 (d6-22: 152±13) compared to T2 (68±5; P<0.0001). The percentage of eaters within a litter did not differ over time between T1 (d10: 26%, d15: 78%, d22: 94%) and T2 (20, 71 and 97%). In conclusion, our results suggest that piglets like to eat a varied diet instead of preferring one feed over the other. Dietary diversity by providing two feeds at the same time different in flavour, size, composition, smell, texture and colour stimulated the feed intake and feed-related exploratory behaviour of suckling pigs more than dietary diversity via novel flavours only, but did not elicit pigs to start eating earlier. Further research is needed to explore the most effective dietary diversity to stimulate early feeding.

Impact of straw particle size and reticulorumen health on the feed sorting behaviour of early lactation dairy cows

Rachael Coon[1], Todd Duffield[2] and Trevor Devries[1]
[1]*University of Guelph, Animal Biosciences, 50 Stone Rd East, Guelph, ON, N1G 2W1, Canada,*
[2]*University of Guelph, Population Medicine, 50 Stone Rd East, Guelph, ON, N1G 2W1, Canada;*
rcoon@uoguelph.ca

The objective of this study was to determine if feed sorting behaviour in early lactation dairy cows varies with straw particle size in a total mixed ration (TMR) and with reticulorumen health. Forty multiparous Holstein cows were enrolled in a trial ~2 wk before calving. Upon calving, cows were fed either 1 of 2 TMR that included 9% wheat straw (dry matter basis) chopped to: (1) 2.54 cm (Short) (n=20); or (2) 5.08 cm (Long) (n=20). Cows remained on their respective treatments for 4 wk. Fresh feed samples were collected 3×/week to determine particle size distribution. Orts samples were collected every 3 d post-calving to determine feed sorting. A particle separator was used to separate feed samples into 4 fractions: long (>19 mm), medium (<19 mm, >8 mm), short (<8 mm, >4 mm), and fine (<4 mm) particles. Sorting was calculated as: actual intake of each particle fraction expressed as a % of its predicted intake. Wireless telemetry boluses were used to measure reticulorumen pH every 15 min, 24 h/d. Cows were classified as having low reticulorumen pH (LPH) if reticulorumen pH fell below 5.8 for an average of >1 h/d throughout the trial (n=15; mean=280 min/d below pH 5.8). If reticulorumen pH fell below pH 5.8 for an average of <1 h/d cows were classified as having normal reticulorumen pH (NPH) (n=25; mean=16 min/d below pH 5.8). Data were summarized by week and analysed in a repeated measures mixed-effect linear regression model. Significance was declared at P≤0.05, and tendencies were reported if 0.05<P≤0.10. Amongst LPH cows, sorting of long particles differed by treatment (P=0.02); LPH cows on the Long treatment (n=9) sorted against those particles (93±3%; P=0.01), while LPH cows on the Short treatment (n=6) tended to sort for those particles (106±3%; P=0.1). NPH cows on both treatments tended (P≤0.1) to sort against the longest particles (96±2%). Sorting of medium particles tended (P=0.06) to differ by treatment for LPH cows; LPH cows on the Long treatment tended to sort against these particles (98±1%; P=0.1), while LPH cows on the Short treatment sorted in favour of them (103±2%; P=0.05). LPH cows sorted short particles differently by treatment (P=0.05); LPH cows on the Long treatment did not sort for or against short particles (101±1%; P=0.5), while LPH cows on the Short treatment sorted in favour of short particles (104±1%; P<0.001). Depending on treatment, LPH cows sorted the fine particles differently (P=0.01); LPH cows on the Long treatment did not sort for or against those particles (102±2%; P=0.2), while LPH cows on the Short treatment sorted against those particles (93±2%; P=0.002). Sorting was consistent across all weeks. These results suggest that those cows who experienced low reticulorumen pH, and were fed a diet with shorter straw particles, sorted that diet to maximize intake of physically-effective fibre to ameliorate the effects of their low reticulorumen pH.

Lowered tail can predict a tail biting outbreak in grower pigs

Helle Pelant Lahrmann[1], Christian Fink Hansen[1], Rick D'Eath[2], Marie Erika Busch[1] and Björn Forkman[3]
[1]SEGES, Danish Pig Research Centre, Axeltorv 3, 1609 Copenhagen, Denmark, [2]Scotland's Rural College, Animal & Veterinary Sciences, West Mains Road, EH9 3JG Edinburgh, United Kingdom, [3]University of Copenhagen, Veterinary and Animal Sciences, Grønnegårdsvej 2, 1870 Frederiksberg C, Denmark; hla@seges.dk

Detecting a tail biting outbreak early is essential in order to stop the unwanted behaviour and thereby the risk of pigs getting a tail injury. Previous studies suggest that tail posture and increasing activity level could be used to predict an upcoming outbreak. Therefore, the aim of the present study was to investigate if differences in tail posture and activity could be detected between upcoming tail biting pens (T-pens) and control pens (C-pens) (>7 days away from an outbreak) using video recordings. The video study included 1,745 pigs in 56 pens (mean ± s.d., 31.2±1.5 pigs per pen). Tails were scored three times weekly between 07:00-14:00 h (for wound freshness, severity and tail length) from weaning until a tail biting outbreak. An outbreak (day 0) was defined as 4 or more pigs having tail damage, regardless of wound freshness. On average 9±4.9 pigs had tail injuries in T-pens (50 pens) on day 0. To eliminate the effect of age and housing, an unaffected pen was identified randomly from the same batch and room for use as a control (C-pen, 25 different pens). A C-pen could therefore be paired up with more than one T-pen. Tail posture and activity was recorded in T-pens and in matched C-pens using scan sampling every 30 min. between 08:00-11:00 h and 17:00-20:00 h on day -3, -2 and -1 prior to the tail biting outbreak in T-pens. At each scan sampling the number of active (standing rather than sitting or lying) pigs and their tail posture (curly or hanging) was recorded. Differences in tail posture and number of active pigs between T-pens and C-pens were analyzed using GLM (PROC MIXED, SAS) including fixed effect of group, day, scan period and age-at-outbreak. Differences on day level were estimated using the Least Square Means statement. Tail posture was not affected by scan period (morning/afternoon) or pigs age-at-outbreak, so these were excluded from the model. No difference in activity was observed between T- and C-pens, but more tails were hanging in T-pens than in C-pens on every recording day (Day -3: 23.0% [CI; 20.9-25.2] vs C-pens: 16.7% [CI; 14.6-18.8], t=4.16, P<0.001; Day -2: 24.8% [CI; 22.7-25.2] vs 15.1% [CI; 13.0-17.3], t=6.3, P<0.001 and Day -1: 33.2% [CI; 31.0-35.3] vs 17.4% [CI; 15.2-19.5], t=10.4, P<0.001). No difference in pigs with the tail hanging was observed between day -3 and -2 in T- pens, but more tails were hanging in T-pens on day -1 compared to day -2 (t=-7.58; P<0.001) and compared to day -3 (t=-9.09; P<0.001). In conclusion, although activity did not increase prior to a tail biting outbreak, tail posture might predict a future tail biting outbreak on pen level: Pens which were about to have a tail biting outbreak within the next 3 days had more hanging tails than control pens, and this difference was greater closer to the outbreak.

Using hen-mounted light sensors to monitor outdoor range use

Stephanie Buijs, Christine Nicol, Francesca Booth, Gemma Richards and John Tarlton
University of Bristol, School of veterinary sciences, Langford house, BS40 5DU Langford, United
Kingdom; stephanie.buijs@bristol.ac.uk

Studying the ranging behaviour of individual hens is very difficult without the use of an accurate automated monitoring system. To assess the accuracy of a novel light-based monitoring system we compared outputs to direct observations of hen location. Fourteen hens from a commercial flock (British Blacktails, 45 weeks old, total flock size 2,000) were equipped with a device that measured and stored light levels each minute. The device was wrapped in brown tape and fitted to the hen's back using elastic loops around the wing base. Attempting to equip hens that varied in their range use, 7 hens were caught on the range and 7 inside the house. Data were collected 2, 3 and 7 days after equipping. Hen location (in/out) was determined by comparing the hen's sensor to four ambient light sensors placed in the brightest areas of the shed. When the reading of a hen-mounted sensor exceeded that of all inside ambient sensors, the hen was considered to be outside. To avoid collecting data when inside and outside light levels were too similar, two additional ambient light sensors were placed outside, and data were discarded when min(out) <1.1×max(in). This resulted in daily data collection starting between 07.11 and 07.45 and ending between 16.20 and 16.35, corresponding with sunrise and sunset for that time of year (November). In addition, 0-30 minutes of data per day when the sun shone through the popholes (around 08:00 or 14:30) had to be discarded. The accuracy of the system was then evaluated by direct observation of the equipped hens (206 minutes in total). One continuous 5 minute observation was conducted per hen/day, between 10.00 and 15.00 on 3 observation days, balancing individuals over different times within those days. Monitoring and observations were in agreement 92% of the time. The monitoring system indicated that hens originally caught outside spent a much greater percentage of the monitored time on the range than those caught inside (median (Q1-Q3): 42% (40-51) vs 6% (3-17), P<0.001, Wilcoxon rank-sum test). We conclude that the system shows great promise as a tool to monitor range use.

Housing laboratory mice from the mouse's point of view

Elin M. Weber, Jerome Geronimo, Jamie Ahloy-Dallaire and Joseph P. Garner
Stanford University, Department of Comparative Medicine, Veterinary Service Center, 287
Campus Drive, Stanford, CA 94305, USA; elweber@stanford.edu

Laboratory mice are housed in small, transparent shoebox cages with bedding and small amounts of nesting material. This housing is fundamentally unnatural; it lacks the complexity of a wild environment, preventing many species-typical behaviours. Present cage designs limit what can be added to a cage, so that current enrichment strategies cannot yield anything close to a naturalistic environment. Therefore, instead of refining present housing systems, our goal was to recreate a mouse habitat. We designed a cage from the mouse's point of view, taking into account that mice are mainly nocturnal, thigmotactic, and excellent climbers. Aggression is a serious welfare problem in mouse husbandry and structural enrichment can increase aggression. However, wild mice use structural complexity to escape from aggressive conspecifics. We therefore provided escape routes throughout the cage to stimulate natural behavioural responses, and ran a pilot study to investigate whether housing mice in the complex cage would induce aggression. For 3Rs reduction, we reused mice from a C57BL/6 background that would otherwise be euthanized. Eight groups of four mice were kept in standard cages (SC) for one week before being transferred to the complex cage (CC). Mice were marked, video recorded for three weeks, and checked for wounds every other day in both SC and CC to detect signs of aggression. Data were analysed in JMP using Repeated measures GLMs. Wounds were found in four of the eight groups. No unwounded group started fighting in CC. However, there was a small increase in wounding ($F_{1,233}$=45.92; P<0.0001) that peaked at 6.5 days in CC ($F_{1,233}$=50.25; P<0.0001) and subsequently healed. We observed fighting for 1 h after wound check for four days/housing using continuous recording. No significant differences were found in mean time spent fighting/hour (SC 2.15 min; CC 1.36 min), or mean fight duration (SC 43.9 sec; CC 34.4 sec). The increase in wounds is consistent with that seen after any cage change, but since animals healed, CC does not appear to present a welfare concern in terms of aggression. Location in CC was noted hourly for the first two days; during this period mice used all parts of the cage, but were mainly observed in the two designated resting areas (72% of observations). As soon as 24 hours after being placed in CC, vertical climbing and jumping were observed, and all groups removed food from the feeder and stored it elsewhere. Further behavioural analysis will reveal information on individual behaviour patterns, to investigate if all mice access all parts of the cage or if some mice monopolize certain areas. This caging system is the first to radically redesign mouse housing taking mouse ecology, sensory biology, and behaviour into consideration. The cage also facilitates visual inspection and cage cleaning without handling the animals. These promising early results potentially provide a completely new way of housing laboratory mice in the future.

Welfare relevance of floor area for single-housed female mink (*Neovison vison*)

María Díez-León[1], Samuel Decker[2], Nora Escribano[3], David Galicia[3], Rupert Palme[4] and Georgia Mason[1]
[1]University of Guelph, 50 Stone Rd E, N1G 2W1 Guelph, Canada, [2]Michigan State Universiy, 474 Shaw Ln, 48824 East Lansing, USA, [3]University of Navarra, Irunlarrea 1, 31008 Pamplona, Spain, [4]University of Veterinary Medicine, Vienna, Veterinärpl. 1, 1210 Vienna, Austria; mdiez@uoguelph.ca

Cage size regulations assume larger floor areas improve welfare, as individuals get more space to perform a diversity of behaviours. Canadian regulations for farmed mink have recently increased minimum floor areas, yet these remain smaller than those mandated in Europe. For single-housed adult females – the subpopulation kept on-farm the longest and thus most at risk of experiencing chronic housing effects – the floor sizes of Canadian cages (CC) are almost half those of European cages (EC); a size difference raising concerns that even these new cages underperform in terms of welfare compared to EC. We tested this hypothesis housing 64 pairs of Black females in either EC (2,613 cm^2) or CC (1,516 cm^2). Each cage housed 2 unrelated females and was provided with an identical nestbox, shelf and 2 enrichments; however, cage size was unavoidably confounded with distance from the aisle. Pairs were split after 3 months (mink being naturally solitary), the remaining 64 females spending 3 more months individually housed in their EC or CC. For 2 weeks at the end of this period we collected data on stereotypic behaviour (SB; % of total activity & % of observations), fearful responses (to handling & in the 'stick test'), and collected samples for faecal glucocorticoid metabolite (FCM) analyses. Half of our subjects were then humanely killed. *Post-mortem*, we estimated degree of tail chewing, extracted and weighed the thymus, spleen, and adrenal glands, and processed their mandibles for fluctuating asymmetry (FA) analyses to assess housing effects on stress physiology, immune function and developmental stress. The remaining females were then given free access to the cage they had not been raised in. After 2-week habituation, their preference for either cage was assessed (as % of time budget, with chance being a distribution the same as the relative sizes of the floor areas). Results indicate no cage effects on SB or tail chewing, but CC-raised females tended to be more fearful (handling: $X^2_{(dfs.1)}$=3.2, P=0.07; stick test: $F_{1,40.7}$=3.1, P=0.08). Stress physiology and immune function were unaffected by cage type (but analyses of FCM & FA are on-going). When a choice of cage was offered, females allocated their overall time equally between EC and CC, but their *active* time was preferentially allocated to the EC (t_{31}=9.7; P<0.001), while the reverse was true for their *inactive* time (t_{31}=6.1; P<0.001). Our results are consistent with mink preferring larger floor areas for active behaviours; yet their choice to rest more often in CC suggests that relative seclusion and ability to monitor their environment (respectively more likely in CC and EC) may have also played a role in mink' preferences. Our data cautiously suggest that females housed for 6 months in EC might experience better welfare as they seem to prefer these larger floor areas, particularly when active, and seem less fearful there. This has implications for future cage size regulations.

Differential effects of enrichment on the subtypes of stereotypic behaviour in mink

Andrea Polanco, María Díez-León and Georgia Mason
University of Guelph, Dept. of Animal Biosciences, 50 Stone Rd East, N1G 2W1, Canada;
apolanco@uoguelph.ca

Stereotypic behaviours (SBs) are common in farmed mink and other captive animals. In carnivores, SBs typically involve whole-body movements (e.g. route-tracing and whole-body bobbing) or head-only movements (e.g. head-twirling). Farmed mink may also repeatedly scratch at the cage walls ('scrabbling'), often in response to the presence of neighbours. Environmental enrichment is commonly used to treat SBs, but typically does not fully eliminate them. We therefore explored whether different subtypes vary in the degree to which they are reduced by enrichment, by assessing enrichment effectiveness on the different SB subtypes of 31 farmed male mink who were individually-housed in standard non-enriched cages from 7 to 12 months of age and then transferred to larger, enriched cages at 14 months old. The amount of change in time spent stereotyping in each subtype was calculated with using two indices: absolute reduction (SB pre-enrichment minus SB post-enrichment) and relative (1 minus SB post-enrichment/SB pre-enrichment) change. Wilcoxon-Signed Rank Tests comparing these metrics for whole-body and scrabbling SB (head-only SB being omitted due to a small n) revealed a significant difference in relative change, whole body SBs being reduced more (whole-body SB: Mdn=1, IQR=0; scrabbling: Mdn=0.93, IQR=0.46; z=-2.43, P<0.05), but not for absolute change. Mink were also scored as to whether or not each subtype was 'cured', with 'yes' indicating that the SB was no longer performed post-enrichment, and 'no' indicating continued performance (even if at reduced levels). A logistic regression, with 'cured' as the dependent variable, and SB subtype and mink ID (a random factor) as the independent variables, revealed a trending effect of SB subtype (x^2=5.32, P=0.07). Specifically, scrabbling was the least likely to be cured by enrichment (P<0.05). Together, the results indicate that scrabbling is less likely to be reduced by enrichment than are other SBs. Thus not all types of mink SB are equally reduced by environmental enrichment, with scrabbling being more resistant to change.

Effects of space allowance and simulated sea transport motion on the behavior of sheep

Ramazan Col[1,2], Grisel Navarro[1] and Clive Phillips[1]
[1]School of Veterinary Sciences, University of Queensland, Centre for Animal Welfare and Ethics, Centre for Animal Welfare and Ethics, School of Veterinary Sciences, University of Queensland, 4343, Gatton QLD, Australia, [2]Faculty of Veterinary Medicine, Selcuk University, Department of Physiology, Department of Physiology, Faculty of Veterinary Medicine, Selcuk University, Campus, 42035 Konya, Turkey; rcol@selcuk.edu.tr

Transporting livestock by ship, especially over long distances, presents significant risks to their welfare. The objective of this study was to assess the effects of space allowance and motion in simulated ship transport on the welfare of sheep. Six sheep (mean weight 25±2 kg) were restrained in pairs in a crate on a programmable platform that generated roll and pitch motion typical of that experienced on board ship for 1 hour/per treatment. Sheep were subjected to movement (regular, irregular motion and control) at Low and High space allowances (SA, 0.26 and 0.52 m^2/sheep) in a replicated Latin square factorial design. Irregular movement was programmed as 30 random pitch and roll movements which varied in amplitude (angle of motion) and period (duration of motion). These were selected by software controlling the movement. Regular movement was the mean of these values, which represented approximately 33% of the recommended maximum tolerance for livestock carriers. Behaviour was continuously video recorded by six cameras around the crate. Low space allowance increased the frequency of pushing, but only in the motion treatments (Low regular 38.7, irregular 30.3, control 14.4 events/h, High regular 4.5, irregular 11.7, control 0 events/h, SED 9.30, P=0.04), and increased time spent being aggressive (Low 4.2, High 0.3 s/h, SED 1.18, P=0.04), affiliative behaviour between the sheep, i.e. one head above the other (Low 43.8, High 6.5, SED 32.89, P=0.02) and standing supported by the crate (Low 449, High 122 s/h, SED 181.4, P<0.0001). Sheep in the Low Irregular treatment spent least time with their head up (124 Vs mean 584 s/h for other treatments, SED 269.5, P=0.008). Irregular movement reduced rumination (864 vs Control 1,776 and Regular 1,208 seconds/hour, Standard Error of Difference 432.7, P=0.02) and increased pawing (20 Vs Control 4 and Irregular 10, SED 13.4, P=0.05) and sniffing (31 Vs Control 3 and Irregular 9 events/h, SED 12.8, P=0.003). Thus there was evidence that low space allowance increased interaction between sheep and that it increased agonistic behaviour more during simulated sea motion than in a static, control situation.

Cooling cows with soakers: effects of flow rate and spray frequency on behavior and physiology

Grazyne Tresoldi[1], Karin E. Schütz[2] and Cassandra B. Tucker[1]
[1]*Center for Animal Welfare, Department of Animal Science, University of California, Davis, One Shields Avenue, Davis CA, 95616, USA,* [2]*AgResearch, Ruakura Research Centre, 10 Bisley Road, Hamilton, 3240, New Zealand; gtresoldi@ucdavis.edu*

Soakers reduce heat load in cattle, however, spray strategies (e.g. flow rate and spray frequency) may affect how cows use this resource and cooling efficiency. Our objective was to evaluate the combined effects of spray frequency (3 min on, 6 min off or 1.5 min on, 3 min off) and flow rates (3.3 or 4.9 l/min) on behavioral and physiological responses to heat load in dairy cows managed in a free-stall barn. In a 2×2 Latin square design, 12 pairs of Holstein cows averaging (±SD) 36±5 kg/d of milk were tested 3 d/treatment when air temperature and relative humidity averaged 23±2 °C and 49±7%, respectively. Water was sprayed at the feedline from 08:15 to 23:30 h (average air temperature ≥19 °C). The overall quantity of water sprayed was not affected by spray frequency; it varied as a function of flow rate (lower: 1,040 vs higher: 1,544 l of water/d). Cows' posture and location within the pen were measured continuously while body temperature was recorded every 3 min 24 h/d. Respiration rates were recorded every 45 min daily from 09:00 to 20:00 h. Data were analyzed with mixed models (SAS) using 24 h averages. Preliminary results indicate that time spent lying down and at the feed bunk were affected by flow rate (P≤0.04), but not by spray frequency (P≥0.35). Under the higher flow rate, cows spent 42 min longer lying down (mean ± SE: 12.1 vs 12.8±0.5 h/d), and used the feed bunk area for 30 min less (5.7 vs 6.2±0.4 h/d). Respiration rate averages were not different among treatments (57 to 59±3 breaths/min, P≥0.22). Although body temperature tended to be reduced when spraying water at higher flow rate (P=0.08) it was within a narrow range, regardless of treatment (38.6 vs 38.7±0.1 °C). Cows adjusted their behavior in response to spray strategies provided. They used the lower flow spray for longer to maintain similar levels of cooling, comparable to those provided by higher flow rates. Spray frequency (when controlling for water volume) did not affect physiological responses or behavior in cattle. Although physiological responses to heat load were not affected, our preliminary behavioral results indicate that other aspects of cow comfort (e.g. less time standing on concrete in front of the feed bunk) are influenced by the cooling strategy used.

Nesting behaviour of broiler breeders

Anne Van Den Oever[1,2,3], Bas Rodenburg[1,2], Liesbeth Bolhuis[2], Lotte Van De Ven[3] and Bas Kemp[2]
[1]Wageningen University, Behavioural Ecology Group, De Elst 1, 6708 WD Wageningen, the Netherlands, [2]Wageningen University, Adaptation Physiology Group, De Elst 1, 6708 WD Wageningen, the Netherlands, [3]Vencomatic Group, Meerheide 200, 5521 DW Eersel, the Netherlands; anna.vandenoever@wur.nl

Broilers have been selected for growth related characteristics, which are negatively correlated to reproductive traits. This genetic background creates challenges in broiler breeders, as the hens do not make optimal use of the nests provided leading to dirty and cracked eggs. Despite the economic importance of this problem, little research has been done on nesting behaviour in broiler breeders. This project aims to investigate which factors are involved in nest site selection of broiler breeder hens and to create practical improvements in housing design. The first experiment focuses on nest design preferences, where three alternatives to a standard nest will be offered to groups of hens. This preference will then be related to the personality traits sociability and fearfulness, measured with density related permanence and tonic immobility tests. The second experiment will compare two genetic strains on leg health and social interactions. The effect of introducing perches at an age of 20 weeks on leg health, measured by scoring for footpad dermatitis and gait, will be investigated. Leg health will furthermore be related to mobility, fertilisation and production. Social interactions amongst females and between females and males are expected to influence nest site selection as well. Dominant females might attract submissive females to certain nests, but can also prohibit submissive females access to the nests. Feral males are known to escort females to suitable nest sites, although it is unknown whether this holds true for domestic chickens. By scoring social interactions and nest site, these possible relations can be studied. A final field experiment will be conducted at a commercial farm, fitted with nests designed according to the preferences of the first experiment and with or without perches as proven optimal according to the second experiment. Productivity, fertilisation, leg health and behaviour will be recorded to measure the performance of this improved system.

Laying hens' behavioural response to wearing automated monitoring equipment

Francesca Booth, Stephanie Buijs, Christine Nicol, Gemma Richards and John Tarlton
University of Bristol, School of veterinary sciences, Langford house, BS40 5DU Langford, United Kingdom; francesca.booth@bristol.ac.uk

Laying hens often respond aggressively to flock members with even minor alterations in appearance, causing altered hens to flee, fight or hide. This can be problematic when equipment for automated behavioural monitoring is attached to hens, as the data supplied will no longer be representative of these individuals' normal time budget. We evaluated the effect of 50 g 'backpacks' on behaviour using direct observation. These 'backpacks' contained two monitoring devices (a light sensor and an accelerometer) and a location device. The equipment was wrapped in brown tape and fitted to the hen's back using elastic loops around the wing base. Fourteen hens (British Blacktails, 45 weeks old) were equipped. These hens were housed on a commercial free-range farm in a flock of 2,000 hens. Behaviour of equipped and non-equipped hens was observed 2, 3 and 7 days after equipping (5 min/bird/day, alternating between equipped and non-equipped hens) and compared using Wilcoxon rank-sum tests. None of the equipped hens were ever observed to flee or fight. Although equipment tended to increase the frequency of pecks received on day 2 and 3 (P=0.08 and 0.05, respectively) compared to non-equipped hens, median values were 0 for both groups and differences in interquartile range were small (0-0 vs 0-0.3 and 0-0 vs 0-0.6 pecks/minute, respectively). Equipped hens generally showed no reaction to these pecks (which included all types of pecks except severe feather pecks). Equipped and non-equipped hens were not found to differ significantly in the proportion of time spent eating/drinking, foraging, perching, preening, sitting, standing, walking or in the nestbox, or in their frequency of body-shaking and pecking other hens (all P-values >0.05 for each day). Although further evaluation in other flocks is required, this system seems to have only a very minor impact on behaviour, confirming its suitability for collecting automated behavioural data.

Effects of stocking density and exercise on incidence of Tibial dyschondroplasia in broilers

Elmira Kabiri[1] and Ahmad Tatar[2]
[1]Behin Toyour Golestan Aviculture Co., Farhang St. Balajade village, Kordkooy, Golestan province, 4883917893, Iran, [2]Ramin Agriculture and Natural Resources University of Khuzestan, Animal Science, Khuzestan, Mollasani, 6341773637, Iran; tatar@ramin.ac.ir

Tibial dyschondroplasia (TD) is the most prevalent skeletal abnormalities associated with rapid growth avian species. It results in deformed bones and lameness. The economic cost of TD as manifested in the rates of mortality, morbidity and condemnation at the processing plant, is enormous as these rates can reach up to 30% of a flock. The exact etiology of TD is unknown. TD is common in fast growing birds, especially in male broilers in which incidence can be high as 30%. Although, rapid growth and other factors like genetic background are contributing factors in the development of TD, they are not the primary factors. Therefore, TD may relate to genetic, nutritional and environmental factors. Placing barriers or ramps within the pen caused an increase in the mass dimensions of broiler breast and leg muscle, indicating that exercise could be encouraged by altering pen design and that such addition to the pen can improve the overall physical condition of the birds. On the other hand, increasing stocking density has been shown to decrease body weight, increase feed conversion and reduce activity level. Reduced activity has also been implicated as a factor involved in the development of leg problems in broiler chickens. In addition to providing birds with exercise, perches in the environment of broiler chickens may improve welfare by providing them with an opportunity to perform a natural behavior and allowing them to exert some control over their environment.

Effects of enriched environments with perches and dust baths on behavioural patterns of slow-growing broiler chickens

Hanh Han Quang, Xuan Nguyen Thi and Ton Vu Dinh
Vietnam National University of Agriculture, Faculty of Animal Science, Trau Quy town, Gia Lam district, Hanoi, Vietnam, 10000, Hanoi, Viet Nam; hanquanghanh1304@gmail.com

The present study was conducted to determine the effects of environmental enrichment with perch and dust bath provision on behavioural patterns of slow-growing broiler chickens. A total of ninety-two local crossbred chickens (Ho × Luong Phuong) were distributed equally into two housing systems, a conventional, totally confined housing system, and an enriched, semi-confined housing system with perch and dust bath availability, with two groups (n=23) per housing system. Chickens' behavioural patterns were observed directly by one assessor by instantaneous scan sampling and the number of birds performing different behaviours were recorded in the morning and afternoon or evening from 6 to 13 weeks of age. The Chi-square analysis with Fisher's test was applied to compare the percentage of birds' behaviours between two treatments by Minitab® 16.0. Chickens at the age of 10 to 13 weeks preferred perching at the resting time during the night with a significant higher percentage of birds perching in the evening than that at daytime (P<0.05). Chickens also showed a low percentage of dust bathing behaviour both in the morning and the afternoon (less than 10%) and it were not significantly affected by age (P>0.05). The percentage of birds that showed the play fighting behaviour was low and not different between two housing systems. However, chickens that had outdoor access were more fearful of human (with a significant higher avoidance distance of birds to human, P<0.05) than those in a totally confinement house. Provision of perches and dust baths helps stimulates the expression of several natural behaviours of broiler chickens.

Farmers' perception of pig aggression and factors limiting welfare improvements

Rachel S.E. Peden[1], Simon P. Turner[1], Irene Camerlink[1] and Laura Boyle[2]

[1]Scotland's Rural College, Animal Behaviour and Welfare, SRUC, Roslin Institute Building, Easter Bush, Roslin, EH25 9RG, United Kingdom, [2]Teagasc, Pig Development Department, Animal & Grassland Research and Innovation Centre, Moorepark, Cork. P61 C997, Ireland; rachel.peden@sruc.ac.uk

Mixing of unfamiliar pigs is common practice in pig husbandry, creating high levels of aggression related to the formation of dominance relationships. Many strategies to reduce aggression have been identified by research but few have been adopted in practice and only by a minority of farmers. As a consequence, aggression remains undiminished in commercial farming. In order to improve welfare by reducing aggression, farmers would have to alter their behaviour and practice regarding pig management. Farmer willingness to change current practice relies strongly on their perception of the situation. Our objective is to approach the welfare problem of pig aggression from the perspective of the industry, focussing on the farmer. UK pig farmers were asked about their management and perception regarding aggression through a postal survey. Of the 167 respondents 73% did not consider aggression at weaning a welfare problem, and 57% did not consider it a problem for grower-finishers. Farmers who did consider aggression a problem for grower-finishers were on average younger than respondents who did not consider aggression a problem (P=0.02). When farmers were asked about pre-weaning co-mingling of adjacent litters, the most researched and one of the most promising methods to reduce aggression, 48% had concerns about the practical management of this technique. If solutions to reduce aggression between pigs were available, 20% of the respondents would consider adopting these methods to manage aggression at weaning, and 37% would do so at the grower/finisher stage. In presenting the results of this research along with data collected in early 2017, we will highlight the contribution of various factors that may limit farmer willingness to change current practice, such as desensitisation, prioritisation of issues, and financial barriers. Knowledge of farmers' perception of aggression, and their perception of proposed strategies to reduce aggression, may help align research questions in this area and find ways to effectively address the issue in a manner that can be adopted by practice.

Behaviour and welfare of cull sows on the day of slaughter – current knowledge and possible ways forward

Mette S. Herskin, Katrine Kop Fogsgaard and Karen Thodberg
Aarhus University, Department of Animal Science, P.O. Box 50, 8830 Tjele, Denmark;
mettes.herskin@anis.au.dk

In recent years, international pig production has been characterised by increasing herd sizes and changes in the slaughter industry towards fewer and larger units. Consequently, the needed transport distances from farm to slaughter are increasing. Across animal groups, this development may have welfare consequences. Especially one group of pigs to be transported – cull animals – may be vulnerable towards transport stress and thus pay a price for the current development in the pig industry. However, despite the large proportion of sows being culled from modern production systems each year, the behaviour and welfare of these animals have received almost no scientific attention. For sows in modern production systems, the typical pre-slaughter logistic chain consists of a series of consecutive potential stressors experienced by the animals on the day of slaughter. Sows may be culled as part of a planned strategy, but for the majority of the animals the decision to cull is a consequence of reduced reproduction, injuries or other health issues. Results from our recent study showed that 40% of cull sows were culled directly at weaning. In Denmark, the cull sows may be loaded directly from their home pen or be mixed with other sows and kept in so-called pick-up pens until they are picked up by the truck. In any case, the sows may be mixed with other sows in a transfer vehicle for biosecurity reasons, and either via very short lasting transit, or a stay for up to two hours, pass this vehicle before being picked up by the commercial truck. Danish law specifies that sows can only be transported for up to 8 hours, whereas sows from other parts of the world may be transported considerably longer, and may pass buying stations or similar before they arrive at the slaughterhouse. Results from our recent study showed that the clinical condition of cull sows, such as number of scratches and measures of dehydration, were worsened by commercial transport for up to 8 h. Among the most important risk factors for the clinical worsening were transport duration and temperature. In this presentation, we review the following phases of the pre-slaughter logistic chain for sows: (1) pre-transport in home pen; (2) on-farm pick-up pens; (3) transfer vehicles; (4) transport; and (5) lairage. Across phases we provide an overview of the available knowledge about behaviour and welfare of sows on the day of slaughter, and discuss possible needs for changes in order to secure the welfare of sows while they pass the pre-slaughter logistic chain.

Clinical consequences of keeping cull sows in on-farm pick-up pens before transport to slaughter

Cathrine Holm, Karen Thodberg and Mette S. Herskin
Aarhus University, Animal Science, P.O. Box 50, 8830, Denmark; holmen_lojt@hotmail.com

In modern pig production up to 50% of the sows are slaughtered each year. Typically, the period from the decision to cull and until a sow is slaughtered includes a stay in an on-farm pick-up pen (hours to weeks), transport (potentially passing buying stations or similar) and lairage. However, even though cull sows are often mentioned as vulnerable towards challenges such as transport, the behaviour and welfare of these animals have received almost no scientific attention. The aim of this study was to examine clinical consequences of keeping cull sows in on-farm pick-up pens before transport to slaughter. The observational study was conducted in 2016 in a private Danish sow herd with loose-housed gestating and lactating crossbred LYD-sows. As part of the farm management, the farmer decided which sows to cull. The study included all cull sows, except if they were recently medicated, severely lame or showed clinical signs of illness. The dataset consisted of 103 sows with a mean parity of 5.2 (range 1-9) and a BCS of 2.9 (range 2-3). At weaning (after 28 days of lactation), all sows to be included in the study were examined clinically, and then walked to the pick-up pens in another part of the same building. Each pick-up pen holding three sows was 4.15×3.25 m with 50% slatted floor and 50% concrete floor. A chained wooden log was provided as manipulable material. The sows had free access to water from a watercup and were fed twice a day on the floor with a total 2.5 FU per pen. The mean temperature was 21.0 °C (range 18-25.6). The sows were kept in the pens until they were picked up by a commercial truck and transported to slaughter. After 24 h in the pick-up pens the clinical condition of the sows showed signs of worsening (examined by Signed Rank Test), as judged by an increased number of scratches (counted on 6 body parts with a maximum number of 15 per part): median 0 (range: 0-22) vs 9 (0-75); $P<0.001$) and increased gait score (0-4): median 0 (range: 0-1) vs 0 (0-2); $P<0.001$) as well as a tendency to an increase in the number of wounds ($P=0.0625$). No relation between the changes in the clinical condition and the temperature during the stay in the pens was found. These results suggest that a 24 h stay in a pick-up pen may be a challenge to the welfare of cull sows and hence potentially affect their fitness for transport. Further studies are needed in order to understand effects of a stay in a pick-up pen as well as other phases of the pre-slaughter logistic chain on welfare during the day of slaughter and to understand possible links with fitness for transport.

Development of an indoor positioning system for automatic monitoring of locomotor behaviour of group-housed sows

Shaojie Zhuang[1,2], *Olga Szczodry*[2,3], *Jarissa Maselyne*[2], *Johan Criel*[4], *Frank Tuyttens*[2,3] *and Annelies Van Nuffel*[2]
[1]*Ghent Univeristy, Faculty of Bioscience Engineering, Coupure links 653, 9000 Gent, Belgium,* [2]*Institute for Agricultural, Fisheries and Food Research (ILVO), Burg. Van Gansberghlaan 92, 9820 Merelbeke, Belgium,* [3]*Ghent University, Faculty of Veterinary Medicine, Salisburylaan 133, 9820 Merelbeke, Belgium,* [4]*Sensolus NV, Kortrijksesteenweg 930, 9000 Ghent, Belgium; olga.szczodry@ilvo.vlaanderen.be*

At times when the numbers of animals per farm are rising, it becomes acknowledged that precision livestock farming (PLF) offers promising tools for improving the welfare of the animals by automatically monitoring, among other variables, their behaviour. Several PLF techniques involve sensors that are capable of continuously collecting data at individual level, thereby offering possibilities for 'management-by-exception'. Our aim was to develop a positioning system that is customizable and accurately tracks animals indoor. For the latter purpose we performed (1) static tests, where position data were collected at 64 predefined locations (approx. 120 s per location, sampling rate 1 Hz, tag height of 1.00 m) and (2) dynamic tests where an investigator walked on a predefined path at a speed of 0~1.50 m/s (1 Hz, 1.00 m tag height). Coordinates of tags were estimated based on tag-to-receiver distances and their accuracy was quantified by their root mean squared distance to the true position. Our preliminary results show that, in static tests, the overall root mean square error (RMSE) of estimated positions is 0.33 m, where 92.2% of test locations have error <0.50 m. The data loss is 3.1%. For the dynamic tests, the RMSE is 0.35 m and 92.6% of the estimates have an error <0.50 m. The data loss is 7.0%. The present results show potential for automatically and continuously collecting position data on animals housed indoor. This offers a wide variety of applications, as for instance, automatic monitoring of locomotor behaviour of sows *via* the estimation of different variables, such as distances covered, walking speed and diversity of areas visited in the pen. Since it has been shown that the latter variables are associated with lameness, the present system has the potential to be used for automatic surveillance of, *inter alia*, lameness status in group-housed gestating sows. This system is currently under development for such an application.

Exploratory behavior in commercial and heritage breed swine

Kristina M. Horback[1,2] and Thomas D. Parsons[2]
[1]University of California Davis, Department of Animal Science, 1 Shields Avenue, 95616 Davis, California, USA, [2]University of Pennsylvania School of Veterinary Medicine, Swine Teaching and Research Center, 382 West Street Road, Kennett Square, PA 19348, USA; thd@vet.upenn.edu

Sows on most commercial farms are the product of intensive breeding programs focused on an ensemble of productivity traits; rarely with any explicit consideration of behavioral responses. This study was designed specifically to see how juvenile behavioral responses to stress-inducing contexts differed between commercial genetics and less intensively selected heritage breed swine. Four neonatal Tamworth litters were fostered onto resident commercial sows (PIC 1050: Landrace-Yorkshire cross) at birth to minimize maternal influences of breed during lactation. Data collection of 14 Tamworth gilts (TAM) and 33 commercial breed gilts (CBG) occurred at processing (1 day old) and 3 weeks old. Measurements recorded include weight (kg), response to restraint via human handling and open field tests. Data were analyzed using SPSS 17.0 with a significance level set of $P<0.05$. Given that the data was not normally-distributed, differences between breeds were calculated using Mann-Whitney U tests and data is presented as median [IQR]. TAM piglets weighed more ($P<0.01$) at birth (1.7[0.3] kg) and at 3 weeks old (6.8[1.3] kg) than the CBG piglets (birth: 1.4[0.3] kg, week 3: 5.1[2.5] kg). TAM piglets received higher scores ($P<0.05$) for the restraint test at 3 weeks old, indicating that this breed showed more resistance to being handled (thrashing and squealing) than CBG. At 3 weeks old, each piglet was given a 5 min open field test (3.4 m^2 arena), beginning with a 30 s rest in a closed off start box. TAM piglets were quicker ($P<0.05$; TAM: 13.4[22.2]s, CBG:25.9[43.4]s) to leave the start box, and rested for a longer duration ($P<0.001$) while in the open field (TAM: 127.2[55.7] s, CBG: 71.1[82.7] s). In contrast, the CGB spent a greater duration nosing the experimental field ($P<0.05$; CBG: 176.0[96.5] s, TAM: 144.5[65.5] s) and produced longer durations ($P<0.05$) of high-frequency squeals while in the open field (CBG: 200.6[69.8] s, TAM: 150.6[89.9] s). While TAM were quicker to leave the start box, they moved and squealed less while in the open field. In comparison, the CBG spent more time nosing the ground and walls of the unfamiliar arena and squealed more. These findings demonstrate that TAM and CBG piglets can exhibit different behavioral responses to stress-induced contexts. Exactly how or if these juvenile behavioral differences translate to different coping strategies in adult sows will require additional study. However these apparent differences could contribute to variation in the ability of individual animals to cope with the potential stressors such as the social complexity found on loose housed sow farms and thus have welfare implications for these animals.

Effect of space allowance and flooring on pregnant ewes' behavior

Stine Grønmo Vik, Olav Øyrehagen and Knut Egil Bøe
Norwegian University of Life Sciences, Department of Animal and Aquacultural Sciences, P.O.
Box 5003, 1432 Ås, Norway; stine.vik@nmbu.no

Space allowance recommendations for pregnant ewes vary considerably. In Norway there is no regulations for space allowance and the majority of ewes are housed on expanded metal flooring with 0.75 to 0.90 m^2/ewe. To investigate the effect of space allowance and type of flooring on the behaviour of ewes, a 3×2 factorial experiment was conducted with 0.75, 1.50 and 2.25 m^2/ewe and straw bedding and expanded metal flooring. Six groups with 8 ewes in each group were exposed to each treatment for 7 days. The ewes were video recorded for 24 hours at the end of each treatment period and general activity, lying position in the pen and social lying position were scored every 15 min. Displacements and aggressive interactions were scored continuously from 10:30 h to 14.30 h. Mean values per group were used as statistical units. Increasing space allowance from 0.75 to 1.50 m^2/ewe resulted in increased lying time, more simultaneous lying and less aggressive behaviour. Sitting was only observed in the 0.75 m^2/ewe treatment and is most likely an effect of difficulties to lay down in such a restricted area. Increasing space allowance further to 2.25 m^2/ewe had no significant effect except from increased distance between ewes when lying. Type of flooring had no significant effect on general activity, but ewes in the straw bedding treatment spent more time lying in the middle of the pen than ewes on expanded metal. In conclusion, increasing space allowance from 0.75 to 1.50 m^2/ewe had positive effects on activity and behavior in pregnant ewes, but increasing space allowance further to 2.25 m^2/ewe had limited effects as well as type of flooring. Hence, recommended space allowance to pregnant ewes should be as close to 1.50 m^2/ewe as possible.

Motivation of dairy cows to enter a milking robot: effects of concentrate type and amount

Julie Cherono Schmidt Henriksen, Martin Riis Weisbjerg and Lene Munksgaard
Aarhus University, Animal Science, Blichers allé 20, 8830 Tjele, Denmark;
juliec.henriksen@anis.au.dk

Milk yield is affected by milking frequency and is thus dependent on cows' motivation to go to the robot voluntarily. If this motivation changes then both yield but also the time budget of the cow may be affected. The aim was to investigate the effects of the type and amount of concentrate on milk yield and behaviour in a 2×2 crossover factorial design. Concentrate S was pelleted including wheat, and the other type, O, was a mix of pelleted concentrate and steamrolled, acidified barley. The amount of concentrate was changed from 3 kg to 6 kg, or opposite. The experimental period was divided into two periods with half of the cows receiving type O and the other half receiving type S during the first period. All cows shifted to the other type of concentrate during the second period. Within period, half of the cows were first offered 3 kg, and the other half 6 kg of concentrate, then after one week, the concentrate offer for each cow was changed to 3 kg or 6 kg over 6 days with 0.5 kg per day. The data were recorded during the following week of constant concentrate offered. The 83 cows (42 Danish Jersey, 41 Danish Holstein) included in the experiment were balanced between treatments according to breed, parity and lactation stage. Data were analysed using a proc mixed model in SAS with treatment, parity, breed and interactions as fixed factors, cow as random effect and days in milk as covariate. Day within cow and week was regarded as repeated observations. Contrasts were made where interactions were significant. The frequency of both number of daily milking (3.0 vs 2.8, P<0.001) and visits in the robot (3.4 vs 3.2, P<0.001) was higher when offered type O vs type S, while there was no effect of changed amount of concentrate on daily milking or visits (P=0.25; P=0.78). In both breeds, energy-corrected milk yield was higher when offered type O (36.7 vs 35.8 kg/day, P=0.001) and when offered 6 kg of concentrate (36.6 vs 35.9 kg/day, P=0.034). For the *ad libitum* offered mixed ration on the feed alley, Danish Jersey had a shorter eating time when offered type O vs type S (155 vs 161 min/day, P=0.026) while Danish Holstein ate for longer time when offered type O (200 vs 194 min/day, P=0.05) in the robot. For both breeds, the eating time of the mixed ration decreased when the cows were offered 6 kg compared to 3 kg of concentrate (182 vs 173 min/day, P<0.001). However, this difference may have minor biological significance. The daily lying time for the Danish Jersey cows did not differ between type of concentrate (O vs S) (10.2 vs 10.2 h/day, P=0.089), but the Danish Holstein spent less time lying when offered type O vs type S (11.1 vs 11.8 h/day, P<0.001). In conclusion, both the type and amount of concentrate affected the daily time budget, but only the type of concentrate affected the motivation to visit the milking robot in this study.

Relationship between daily lying time and weight during lactation in Holstein and Jersey cows

Lene Munksgaard[1], Julie Cherono Schmidt Henriksen[1], Martin Riis Weisbjerg[1] and Peter Løvendahl[2]

[1]Aarhus University, Animals Science, Blichers Allé 20, Postbox 50, 8830 Tjele, Denmark, [2]Aarhus University, Molecular Biology and Genetics, Blichers Allé 20, Postbox 50, 8830 Tjele, Denmark; lene.munksgaard@anis.au.dk

Sufficient lying time has high priority in dairy cows, and devices for automatic recording of lying behaviour on commercial farms is now available. New results suggest that daily lying time changes dramatically during lactation. However, to take advantage of information about lying behaviour we need to know more about how breed, stage of lactation and parity affects lying behaviour. We hypothesized that the relatively large changes in weight during early lactation relate to lying behaviour. Therefore we investigated the relation between daily lying time and weight changes during lactation in Holstein (69 first parity, 87 with average parity 2.77±0.95) and Jersey cows (52 first parity, 86 with average parity 3.05±1.09) kept in a loose housing system with cubicles and automatic milking. Live weight was recorded whenever the cows were entering the robot and an average daily weight was calculated. Lying time was recorded with a leg based sensor (AfiTagII, AfiMilk, Israel). Data from day 4 to day 250 in lactation were included in the analysis. A Wilmink function was fitted to data based on each breed and parity, and slopes of the individual animals were included as random effects. Correlations between random solutions for lying time and weight were estimated. Throughout lactation Jersey cows were spending less time on lying than Holstein (638 vs 707 minutes / day, $P<0.0001$). In both breeds the profile of lying time during lactation was substantially different between first and later parity cows. In first parity lying time increased until day 100 in Jersey and day 125 in Holstein cows, and stayed afterwards reasonably constant until day 250. In older cows the lying time decreased reaching a nadir approximately 30 days in milk. In contrast to first parity, lying time for older cows increased steadily after nadir until day 250. In both first parity and older cows the weight decreased in early lactation and then increased. Thus the weight change and lying time followed the same pattern over lactation in older cows, but not in first parity cow. Pearsons correlations between weight change and change in lying time estimated by the Wilmink function were weak (numerical <0.25) and non-significant in both first parity and older cows and in both parts of the lactation. In conclusion, lying time in older cows follow the same pattern as the energy balance with a drop during the first month of lactation, however we found only weak correlation between weight change and change in lying time at the individual cow level.

Effects of stocking density on bedding quality and indicators of cow welfare in hybrid off-paddock facilities

Cheryl O'Connor, Suzanne Dowling and Jim Webster
AgResearch Ltd, Ruakura Research Centre, Private Bag 3123, Hamilton 3240, New Zealand;
cheryl.oconnor@agresearch.co.nz

The dairy industry in New Zealand is pasture-based, but hybrid off-paddock facilities where cows spend part, or all, of the day are becoming more commonplace. It is important to define good practice and to understand the welfare risks associated with these hybrid off-paddock systems. One of the key knowledge gaps is the effect of stocking density on bedding quality and lying time within an off-paddock environment. One hundred and thirty-six non-lactating dairy cows were divided into replicated groups of two commonly-used stocking densities of 28 (7.85 m^2/cow) and 39 (5.6 m^2/cow) cows. Cows were kept on a pad of post-peelings for 18 h/day, with 6 h/day on pasture, for 30 days during winter. Lying times were recorded continuously for all cows using Onset Pendant G data loggers and analysed using a mixed model smoother. Gait, dirtiness and body condition scores were recorded once a week. Bedding quality measures, moisture content, a visual description of the ratio of post-peelings to muck and a farmer-friendly 'Gumboot score' (the depth that a person standing in boots sinks into the muck, on a scale of 0-3), were taken weekly from four 25×40 cm quadrat samples across each pad. Bedding quality and cow measurement data were analysed using a repeated measures mixed model in Genstat. Daily lying time declined during the trial from 11.7 h/day to 4.7 h/day (SED=0.19; P<0.001). There was heavy rainfall in the last week of the trial, and concurrent troughs in daily lying time. On rainy days, in particular, most of the herd lay for less than the industry minimum recommendation of 8 hours/day. The lying pattern of the herd also changed with more cows lying down sooner, and for longer on pasture. There were no significant effects of stocking density on any of the measures except the mean cow dirtiness score at week 3, and lower rear limb dirtiness overall, which were higher at the higher stocking density. The declines in lying time and bedding quality tended to be earlier and greater at the higher stocking density. Over the 30 days of the trial, bedding moisture content increased from 64 to 86% (P<0.001). The 'gumboot score' and visual percentage muck score also increased significantly over the trial period (P<0.001). The moisture content was over 80% for all groups when daily lying times were less than the recommended 8 hours/day. A combination of visual and gumboot scores provided a good indication of moisture level. When cows are kept in hybrid off-paddock facilities, we recommend that farmers use bedding moisture content indicators and lying behaviour patterns at pasture as convenient comfort indicators for cow welfare.

The effect different tie-rail placements have on ruminating and feeding behavior of dairy cows housed in tie-stalls

Jessica St John[1], Elsa Vasseur[1] and Jeff Rushen[2]
[1]McGill, Animal Science, 21111 Lakeshore Road, Ste. Anne de Bellevue, Quebec, H9X3V9, Canada, [2]University of British Columbia, Dairy Education and Research Centre, 6947 #7 Lougheed Highway, Agassiz, BC, V0M1A0, Canada; jessica.st-john@mail.mcgill.ca

Although the majority of dairy farms in Canada utilize tie-stall housing there is little information on cow comfort based on experimental studies for cows housed in tie-stall barns. From epidemiological studies, potential inadequacies for the current recommendation were observed for tie-rail placement which may lead to reduced cow welfare. With the implementation of animal care assessments through certification programs there is also a need to test and potentially develop new recommendations to help farmers improve dairy cow welfare. The objective of this study was to develop new recommendations for tie-rail placement combining both vertical and horizontal positions of the tie-rail. Four treatments were tested: two new tie-rail positions that follow the neck line of the cows (Neckline1 and Neckline2), current recommendation as a control (Current Recommendation), and the tie-rail position most commonly found on farm (Common on Farm). Forty-eight cows (12 cows per treatment) were divided into two separate start dates and were blocked by number and stage of lactation then randomly allocated to a treatment over a 10-week period. Ear-mounted data loggers were used to continuously recorded rumination and feeding time and the loggers reported the data as percentage of time spent ruminating/feeding per hour. Differences between/within treatments overtime were analyzed using a mixed model where treatment, week, start date, and block were fixed effects and cow was a random effect. Multiple-comparisons with a Scheffé adjustment was performed to analyze differences among the main effects of treatment and week. On average cows housed in the Neckline1 treatment were found to spend 8.5% less time per hour feeding than cows in the Current Recommendation treatment ($P<0.01$), and up to 4.2% less time per hour feeding than cows housed in the Common on Farm and Neckline2 treatments ($P<0.01$). Cows in the Current Recommendation treatment spent up to 5.2% more time per hour feeding than cows in the Common on Farm and Neckline2 treatments ($P<0.01$). The average percentage of time per hour spent ruminating for week 9 ($40.1\pm0.86\%/h$) was significantly different from week 5 ($44.4\pm0.86\%/h$; $P<0.01$), week 6 ($43.5\pm0.86\%/h$; $P<0.05$) and week 7 ($43.0\pm0.86\%/h$; $P<0.05$). A difference in average time spent ruminating was also found between week 8 ($41.2\pm0.86\%/h$) and week 5 ($44.4\pm0.86\%/h$; $P<0.05$). Results suggest that tie-rail placement may have an effect on time spent feeding but not time spent ruminating.

Characterizing changes in feeding behaviour of lactating dairy cows during the peri-oestrus period

Hawar M. Zebari, S. Mark Rutter and Emma C.L. Bleach
Harper Adams University, Department of Animal Production Welfare and Veterinary Services, Edgmond, TF10 8NB Shropshire, United Kingdom; hzebari@harper-adams.ac.uk

The normal time budgets of dairy cows are influenced by oestrus, with cows spending less time resting and eating but more time walking. Cows have been shown to spend approximately 21% less time feeding where the day of oestrus is assumed to be the day of successful artificial insemination. The objective of the present study was to determine whether dry matter intake, feeding duration and the number of visits to feed were affected by behavioural and silent oestrus in lactating dairy cows. Thirty Holstein Friesian cows were used for the study (initial bodyweight 637.2±59.9 kg (mean ± SEM) and 29.6±6.2 days postpartum) at the Harper Adams University dairy unit. Cows were housed in a free-stall barn with 34 cubicles (2.7×1.2 m) and were observed for signs of oestrus during three 30 minute periods (07:30, 12:30 and 19:30) daily. Milk samples (40 ml) were collected on Monday, Wednesday and Friday afternoon and stored at 4 °C until analysis for progesterone concentration by enzyme immunoassay. A cow was considered in behavioural oestrus when points scored at three consecutive observations was >50 and progesterone concentrations were <2 ng/ml, followed by an increase to >15 ng/ml. A cow was considered in silent oestrus when the score was <50 points, during oestrus as defined by milk progesterone profile. Daily dry matter intakes, feeding duration and the number of visits to feed by cows were recorded by a Roughage Intake Control system (Insentec B.V., Marknesse, the Netherlands). The data were analysed by repeated measures ANOVA (GenStat 17th edition). The Animals Experimental Committee of Harper Adams University gave ethical approval for the study. On the day of oestrus, dry matter intake (19.8±0.41 kg/d), feeding duration (2.6±0.09 h/d) and the number of visits to feed (25.3±1.26 visits/d) were significantly lower (P<0.001) compared to 3 days before (22.4±0.41 kg/d, 3.4±0.17 h/d, 28±1.8 visits/d) and 3 days after (22.6±0.5 kg/d, 3.2±0.12 h/d, 27.9±1.23 visits/d). During behavioural oestrus, dry matter intakes, feeding duration and the number of visits to feed were 19.8±0.41 kg/day, 2.6±0.09 h/d and 25.3±1.26 visits/d, respectively compared to silent oestrus, 20.0±0.61 kg/day, 2.9±0.15 h/d and 26.1±1.5 visits per day, respectively. On the day of silent oestrus, only duration of feeding was significantly reduced (P<0.03) compared to one day before (3.4±0.2 h/d) and one day after (3.5±0.2 h/d) oestrus. There was no significant interaction between oestrus expression and time with regard to dry matter intake, feeding duration and number of visits to feed. In conclusion, although dry matter intake, feeding duration and the number of visits to feed were reduced by behavioural oestrus, only feeding duration was significantly lower during silent oestrus.

Inclusion of feeding behaviour to improve the accuracy of dry matter intake prediction in dairy cows

Guilherme Amorim Franchi, Natascha Selje-Aßmann and Uta Dickhoefer
University of Hohenheim, Institute of Agricultural Sciences in the Tropics, Section of Animal Nutrition and Rangeland Management (490i), Fruwirthstraße 31, 70599 Stuttgart, Germany; inst490i@uni-hohenheim.de

The inclusion of feeding behaviour parameters into models to predict dry matter intake (DMI) may enhance their accuracy. In this regard, this study aimed at evaluating the accuracy of 16 models published in the literature to predict DMI of dairy cows and at developing a model that additionally incorporates feeding behaviour parameters. Lactating Holstein-Friesian cows, with a mean initial live weight and milk yield of 708.3 kg (standard deviation [S.D.] 81.3) and 27.9 kg/d (S.D. 10.9), respectively, were fed total mixed rations mainly composed by grass silage, maize silage, and a concentrate mixture (forage:concentrate ratios of 67:33 or 70:30). The animals were equipped with noseband pressure sensors to monitor jaw movements and quantify eating and ruminating activities during seven consecutive days per cow, in average (S.D. 1.6). The DMI of individual cows (mean 22.1 kg/d, S.D. 4.4) was recorded by automatic weighing troughs. In summary, a dataset of 336 individual cow observations was obtained and posteriorly divided into two subsets, n_1=279 used to calibrate the new model and n_2=57 used for model validation. Estimates of DMI from the 16 models were compared to the actual DMI through paired t test. Using a stepwise selection procedure (PROC GLMSELECT and PROC REG, SAS 9.4) with α=0.05, a new model including parameters of feeding behaviour as well as those used in published prediction equations was created. A linear model regression and a paired t test were performed to compare the DMI estimated by the new model to the actual DMI of cows of dataset n_2. Finally, the role of the included feeding behaviour parameters in enhancing the DMI prediction accuracy was assessed by removing these parameters from the new model, regressing the new estimates of DMI against the actual DMI, and contrasting the resultant R^2 with the one from the first regression analysis. Five models successfully predicted DMI (P>0.05), and their biases ranged from -0.38 to 0.16 kg/d with mean-square prediction errors varying from 6.57 to 17.8 kg^2/d. The new model was DMI [kg/d] = 6.65 + (0.001 × *Age [d]*) – (0.00001 × *Lactation length^2* [d]) + (3.17 × *Lactation length$^{0.178}$* [weeks]) – (0.598 × *Mouth Width [cm]*) + (12.04 × *Milk Protein Yield [kg/d]*) + (0.000018 × *Liveweight^2 [kg]*) + (0.012 × *Eating Time [min/d]*) + (0.178 × *Ruminating Chews [n/min]*) (R^2=0.81, bias = 0.1 kg/d). Estimates derived from this model were highly correlated with (R^2=0.75) and did not differ from (P>0.05, bias = -0.03 kg/d) actual DMI of cows in dataset n_2. The addition of feeding behaviour parameters and of mouth width as a proxy of bite size improved the accuracy of DMI prediction by 11%, indicating that it may considerably improve the accuracy of DMI predictions in dairy cows. Nevertheless, the proposed model should be validated in trials with more animals and for different dairy farming systems and diets.

Validation of the ability of a 3D pedometer to accurately determine number of steps in dairy cows when housed in tie-stall

Elise Shepley[1], Marianne Berthelot[2] and Elsa Vasseur[1]
[1]McGill University, Animal Science, 21111 Lakeshore Rd, Ste-Anne-de-Bellevue, QC, H9X 3V9, Canada, [2]Agrocampus Ouest, 65 Rue de Saint-Brieuc, 35000 Rennes, France; elise.shepley@mail.mcgill.ca

Increasing importance is being placed on the ability of dairy cows to have access to exercise. Automated methods for monitoring activity, such as 3D pedometers, provide the means for researchers and producers to quantify the amount of exercise a cow's environment provides. However, such technologies are not often validated for use in all dairy housing systems with the most frequent gaps being in tie-stall barns – the predominant housing system in Canada. The objectives of the current study were: (1) to determine the ability of the IceTag™ 3D pedometer to accurately measure step data for cows in tie-stall; and (2) to determine whether the leg on which the pedometer is mounted impacts step data. Twenty cows were randomly selected and recorded for six hours each during three separate 2-hour periods. Hours of recording were selected based on when cows in the barn were most active to maximize the number of steps compared. Prior to recording, cows were equipped with a pedometer on each rear leg. Exact start time of the video recordings were documented and lying time was used to confirm that video and pedometer minutes were the same. An ethogram defining step activity was developed by comparing second by second data on the video recording to the pedometer readouts. Two observers were trained to observe for step activity with inter-observer reliability calculated as Kw=0.82. Intra-observer repeatability was calculated as Kw=0.88 and 0.86 for observer 1 and 2, respectively. Video recordings for the 20 cows were then viewed and the total number of steps per minute were recorded for each of the six hours of recordings. Hourly averages for right and left leg data were analyzed separately using a multivariate mixed model with number of steps as a dependent variable and technology and time as fixed effects. This allowed for computation of the correlation between pedometer and video step data. The same model was used to determine correlation between left and right leg step data with number of steps as the dependent variable and leg and time as the fixed effects. The analysis of the video vs pedometer data yielded a high overall correlation for both the left (r=0.93) and right (r=0.95) leg. Furthermore, total number of steps between the left and right leg were found to also be well correlated (r=0.80). These results indicate that the IceTag™ 3D pedometers were reliable for calculating step activity in tie-stall housed dairy cows and can be mounted on either leg with similar step results. The accuracy of this pedometer technology to measure step activity in tied dairy cows provides us with a useful automated tool to determine how to better address exercise needs of dairy cows in tie-stalls.

Lameness in grazing dairy herds: cow level risk factors

José A. Bran[1], Rolnei R. Daros[2], Marina Von Keyserlingk[2], Stephen Leblanc[3] and Maria José Hötzel[1]
[1]*Universidade Federal de Santa Catarina, Departamento de Zootecnia e Des. Rural, Rodovia Admar Gonzaga, 1346, 88034000, Florianópolis, Brazil,* [2]*University of British Columbia, Animal Welfare Program, Faculty of Land and Food Systems, 2357 Main Mall, BC V6T 1Z4 Vancouver, Canada,* [3]*University of Guelph, Population Medicine, Ontario Veterinary College, 50 Stone Road E., N1G 2W1 Guelph, Canada; maria.j.hotzel@ufsc.br*

Lameness reduces feed intake, increases reproduction failure and risk of culling in dairy cows. Little information is available on lameness in grazing cows. This study aimed to assess the occurrence and cow level associated factors to lameness in smallholder grazing dairy herds. Farms (n=41) in the south of Brazil were visited twice, and all lactating cows (n=1,110) were gait scored (1-5 points; lame: $\geq$3). The effects of breed (b), parity (P), presence of superficial hoof abnormalities (ha: scissor claw, inter-digital skin hyperplasia, digital dermatitis, horn fissures) and body condition score (bcs) were assessed. Mixed models (logistic regression using herd as random effect) were fitted for incident ('I' = new case), chronic ('C' = lame on two visits) and recovered ('R' = after the first visit) cases of lameness. Lameness incidence was 28.6%. The ratio of new-cases:recovered-cases was 2, showing progression of the problem, despite the recovery of some cows. The ratio healthy-cases:new-cases was 2.5 and the ratio healthy-cases:chronic-cases was 3. Probability of I cases increased with Holstein breed (OR 3.3), higher parity (OR P 2-3: 2.4; OR P>3: 6.4), bcs$\leq$3 on first visit (OR 2.1), any ha (OR 2.4). Probability of C cases increased with Holstein (OR 7) or crossbred cow (OR 11), old cows (OR p 2-3 8; OR P>3 54), any ha (OR 2.6). Being a Jersey or crossbreed (OR 3.2) or a young (OR p 1-2 3.6) cow increased the probability of recover. The relationship between breed and lameness might not be causal, as breed may be confounded with other factors (e.g. milk yield, feeding practices). The risk factors associated with lameness on smallholder dairy farms seems similar to that previously reported in different dairy cattle housing systems. We also encourage future work to disentangle the interaction between breed and grazing systems.

Grouping, synchrony and comfort related behaviors of dairy cows are affected by the presence of trees in the paddocks

José A. Bran, Thomás L. Ferreira, Luã Veiga, Sérgio A.F. Quadros and Luiz C.P. Machado-Filho
Universidade Federal de Santa Catarina, Departamento de Zootecnia e Des. Rural, Rodovia Admar Gonzaga, 1346, 88034000, Florianópolis, SC, Brazil; pinheiro.machado@ufsc.br

The aim of this study was to assess the effect of tree density on the behaviors of a single herd of 16 lactating dairy cows. The herd was observed over 6 days (8 h/d) in 6 outdoor paddocks: 3 with high (H: 200-300 m^2 of shadow) and 3 with low (L: 20-60 m^2 of shadow) tree density. The lying behavior, location (shade or sun), and a synchrony index (SI) of behaviors (grazing, rumination) were recorded via live observation; these behaviors were sampled every 10 min. Grooming and browsing in the trees were sampled continuously. Lying, grooming and browsing were used as indicators of cow comfort behaviors. Weather information was collected. Respiratory rate, rectal temperature and milk yield were used as indicators of each cow's thermoregulatory state. Chi-square test was used to compare the frequency of behaviors and Kruskal-Wallis to compare clustering behavior. A high proportion of cows were grouped in the shadow in the H paddocks (median: 0.63, 0.60, 0.47; L paddocks median: 0.26, 006, 0.06, P<0.05). The daily temperature and humidity index (THI) remained within the thermal neutral zone, and we found no effect on the physiological indicators of thermoregulation. The SI was negatively correlated with the THI in all the paddocks (L: -0.92; H: -0.68) indicating that the herd became less synchronous at higher THI. The frequency of grooming with trees (158 vs 55) and browsing (75 vs 11) was higher in H than in L paddocks (P<0.05). The number of lying bouts and rumination while lying 7.4 and 8 times higher in H than in L paddocks (P<0.05). Increasing the number of trees in paddocks may modify the grouping and synchrony of the herd, as well as may influence cow comfort behaviors by occupational, physical and sensitive environmental enrichment

The accuracy of accelerometer-based, leg-mounted sensors for measuring dairy cow locomotion and lying behaviour at pasture

Gemma L. Charlton, Carrie Gauld, Emma C.L. Bleach and S. Mark Rutter
Harper Adams University, Animal Production, Welfare and Veterinary Sciences, Shropshire, TF10 8NB, United Kingdom; gcharlton@harper-adams.ac.uk

Lying time is an important measure of cow comfort, and when used in conjunction with other measures it can be used to detect health and welfare problems. Accelerometers can be used to automatically record lying behaviour, and this approach has been previously validated in housed cattle. As previous research shows that accelerometers may overestimate the number of lying bouts in cows at pasture, the objectives of this study were firstly to validate the use of IceQube accelerometers (IceRobotics Ltd., Edinburgh, UK) for recording lying time, number of steps and transitions between lying and standing and vice versa of dairy cows at pasture. The second objective was to determine whether there are any differences in behaviour, as recorded by accelerometers on the front vs back leg. As part of a larger study at Harper Adams University, Holstein Friesian cows, at various stages of lactation were fitted with IceQube accelerometers; one on the back left leg (BL) and one on the front left leg (FL) for 24 months. The sensors record lying and standing duration, frequency and duration of lying and standing bouts and step count. During the summer of 2015, 48 cows were manually observed for 2 h at pasture. Posture was recorded (lying, standing, walking), as well as all leg movements of the front and back left leg. Movements included number of steps, leg lifts and kicks. During the manual observations the cows spent, on average, 42.5 mins ($\pm$3.08; range: 0-120 mins) lying down, they took 332.9 steps ($\pm$19.42; range: 0-792 steps) and transitioned 1.2 times ($\pm$0.11; range 0-5 transitions) between lying and standing and vice versa. Linear regressions (Genstat, 17th edition, VSN International Ltd., UK) revealed strong, positive correlations between the BL IceQube and manual observations for total lying time (r^2=99.5; P<0.001), total number of steps (r^2=85.6; P<0.001) and the total number of transitions (r^2=76.1; P<0.001). There were positive correlations between FL IceQube and manual observation for total lying time (r^2=89.4; P<0.001), total number of steps (r^2=68.2; P<0.001) and the total number of transitions (r^2=44.6; P<0.001), however these were not as strongly correlated as BL. These findings suggest that IceQube accelerometers on the back leg give an accurate record of lying, stepping and transition behaviour of dairy cows at pasture. The IceQube was designed and calibrated for hind leg use, and this experiment verifies that front leg use would reduce performance.

Motivation of dairy cows to access a mechanical brush

Emilie A. McConnachie[1], Alexander J. Thompson[1], Anne-Marieke C. Smid[1], Marek A. Gaworski[2], Daniel M. Weary[1] and Marina A.G. Von Keyserlingk[1]
[1]University of British Columbia, 2357 Main Mall, V6T 1Z4, Canada, [2]Warsaw University of Life Sciences, Nowoursynowska 164, 02-787, Poland; marina.vonkeyserlingk@ubc.ca

Mechanical brushes are increasingly installed in dairy barns, but little is known about how highly cows value brush use. Motivation testing can be used to assess the importance of different resources for animals. One way to assess motivation is to train animals to push open a weighted-gate for access to the resource, then increasing the weight required to open the gate over time. The more weight the animal is willing to push to access the resource, the more important we infer that resource to be. Our aim was to assess motivation of dairy cows to access a mechanical brush. Over two weeks, 10 Holstein cows were trained to push open a weighted-gate. Motivation to access the mechanical brush was compared with that to access fresh feed (positive control) and an empty pen (negative control), and finally motivation to access the mechanical brush was retested at the end of the experiment. The maximum weight that cows were willing to push was analyzed using Kaplan-Meier survival plots. This analysis showed no difference in motivation to access the mechanical brush and fresh feed (P=0.46; P=0.52; mean max weight ± SEM: brush = 48±7.4 kg and TMR = 56±8.8 kg), but cows were less motivated to access the empty pen (mean max weight = 8±4.4 kg) than the they were to access the either the brush or the feed (P<0.001). We conclude that dairy cows are highly motivated to access a mechanical brush.

Association of cow- and farm-level factors with lying behaviour and risk of elevated somatic cell count

Ivelisse Robles[1], David Kelton[2], Herman Barkema[3], Gregory Keefe[4], Jean-Philippe Roy[5], Marina Von Keyserlingk[6] and Trevor Devries[1]

[1]*University of Guelph, Animal Biosciences, Guelph, ON N1G 2W1, Canada,* [2]*University of Guelph, Population Medicine, Guelph, ON N1G 2W1, Canada,* [3]*University of Calgary, Faculty of Veterinary Medicine, Calgary, AB T2N 1N4, Canada,* [4]*University of Prince Edward Island, Atlantic Veterinary College, Charlottetown, PE C1A 4P3, Canada,* [5]*Université de Montréal, Faculté de Médecine Vétérinaire, Montréal, QC H3T 1J4, Canada,* [6]*University of British Columbia, Animal Welfare Program, Vancouver, BC V6T 1Z4, Canada; irobles@uoguelph.ca*

The objective of this study was to associate cow- and farm-level factors with the lying behaviour of lactating dairy cows and their risk for elevated somatic cell count (SCC). Cows from 18 commercial free-stall dairy herds in Ontario, Canada were enrolled in a longitudinal study. Four hundred cows in total were selected for the study based on days in milk (<120 d), absence of mastitis treatment in the last 3 mo, and SCC (<100,000 cells/ml). Data on SCC were collected through regular milk testing (~5-wk intervals). The study began within 7 d after a milk test and continued until 3 tests were completed (~105 d), for a total of 3-observation periods/cow. Elevated SCC (eSCC) was used as an indicator of subclinical mastitis. An incident of eSCC was defined as a cow having a SCC>200,000 cells/ml at the end of a period when SCC was <100,000 cells/ml at the beginning of that period. Lying behaviour was recorded for 6 d after each milk sampling, using electronic data loggers. At the end of each recording period, cow body condition score (BCS) was recorded using a 5-point numerical rating scale (1=thin to 5=fat). Details of barn design, stocking density, and herd management were collected. Stall cleanliness was assessed with a 1-m2 metal grid, containing 88 squares, centered between stall partitions of every 10th stall in each farm, and then counting the squares containing visible urine and/or fecal matter. Factors associated with cow lying behaviour were analyzed using multivariable mixed-effect linear regression models. Cows averaged 656±3.3 min/d lying time. On average, under-conditioned cows (BCS≤2.5) spent 37 min/d less (P=0.03) time lying down than over-conditioned cows (BCS≥4.0). On average, cows tended to spend 35.8 min/d more time (P=0.06) lying down in deep-bedded vs mattress-based stalls. Mean proportion of soiled squares/stall was 20.1±0.50%. As the proportion of soiled squares/stall increased across farms, cow lying time decreased (P=0.003); the model predicted a difference in daily lying time of ~80 min/d between the farms with the cleanest stalls as compared to those with the dirtiest stalls. Factors associated with risk of an eSCC were analyzed using a multivariable mixed-effect logistic regression model. Over the study period, 50 eSCC were detected, resulting in an incidence rate of 0.45 eSCC/cow-year at risk. No factors were associated with risk of new eSCC; this is likely due to low frequency of new cases of eSCC across the study period. Overall, these results confirm our knowledge, that on commercial farms, cows prefer to lie down in cleaner and more comfortable environments.

Who's stressed: nonlinear measures of heart rate variability may provide new clues for evaluating the swine stress response

Christopher J. Byrd[1], Jay S. Johnson[2] and Donald C. Lay, Jr.[2]
[1]Purdue University, Department of Animal Sciences, 125 S. Russell St., West Lafayette, IN 47907, USA, [2]USDA-ARS, Livestock Behavior Research Unit, 125 S. Russell St. Rm. 216, West Lafayette, IN 47907, USA; byrd17@purdue.edu

Heart rate variability (HRV) is a well-known proxy for assessing autonomic nervous system function in swine. Several traditional linear HRV indices such as inter-beat interval length (R-R), and standard deviation of inter-beat interval length (long term, SDNN; short term, RMSSD) provide valuable data for understanding an animal's response to common on-farm stressors. Nonlinear HRV indices provide new information by quantifying the complexity of heart rate fluctuations but have not been used in swine welfare studies. The study objective was to evaluate whether nonlinear indices could be used to detect differences in HRV that may go undetected by linear indices following an acute heat stress episode, similar to what swine can experience during summer in the U.S. Methods were approved by the Animal Care and Use Committee (#1604001402). Eight pigs, 13 to 15 weeks old (6 gilts, 2 barrows; 49±1.5 kg) were individually housed in thermoneutral (TN) conditions (22.4±0.1 °C). Each pig was orally administered an ingestible temperature sensor 16 h prior to testing. Heart rate data were collected under TN conditions in their home pens for 1 h (Phase 1). Following Phase 1, pigs were moved to a separate room and exposed to acute HS (Phase 2; 40.3±0.05 °C). When pig body temperature reached 40.5 °C, heart rate was collected for 1 h, or until body temperature reached 41.5 °C. Immediately prior to- and following Phase 2, blood was collected for analysis of cardiac troponin 1 (cTn1), a metabolite indicative of myocardial stress. Heart rate data were analyzed and corrected using previously published error-correction guidelines. Behavioral data (active/inactive and posture) were collected continuously from video. Datasets of 1000 beats with less than 5% error during inactivity were used for calculation of HRV indices. Data were transformed when needed to meet assumptions of normality and homogeneity of variance. The effect of body weight, breed, sex, and phase (1 or 2) on individual linear (heart rate, R-R, SDNN, and RMSSD) and nonlinear (sample entropy, SD1/SD2, shannon entropy, recurrence rate) HRV indices, cTn1 levels, and time spent active were tested using repeated measures mixed model ANOVAs. Comparisons were adjusted using the Tukey test. Sample entropy was higher during Phase 1, indicating greater HRV complexity compared to phase 2 (1.3±0.1 vs 0.8±0.1; P=0.03). Phase was not associated with any of the remaining HRV indices. Levels of cTn1 were lower in Phase 1 compared to Phase 2 (0.05±0.01 vs 0.10±0.02; P=0.05). Activity levels in Phase 1 (0.26±0.05) were not different from Phase 2 (0.15±0.05; P>0.05). Breed, sex, and weight were not associated with HRV indices, cTn1 concentration, or activity levels. In summary, none of the linear measures of HRV indicated pigs were distressed. In contrast, the non-linear measure, sample entropy, did show the pigs were distressed and may add value to HRV studies focused on improving swine welfare.

Effects of space allowance and simulated sea transport motion on the behavior and heart rate of sheep

Grisel Navarro
Univeristy of Queensland, Centre for Animal Welfare and Ethics, School of Veterinary Science Building 8134, Gatton Campus University of Queensland, 4343, Australia; grisel.navarrootarola@uq.net.au

We have previously demonstrated that a low space allowance increased interactions between sheep in simulated ship transport, both negative, i.e. pushing, and positive, i.e. affiliative behaviour, where one sheep puts its head beneath the other's head. The net impact on stress to the animals is unclear, therefore we additionally monitored heart rate and its variability. Nine sheep were exposed to three stocking densities (High H, representing the Australian shipping standard for these sheep, Medium M and Low L, 0.78, 0.92 and 1.04 m2/sheep, respectively) in triplets in a replicated Latin square factorial design. Sheep were placed in a crate on a programmable platform that generated roll and pitch motion typical of that experienced on board ship for 1 hour/per treatment. Motions were either irregular (I), programmed as 30 randomly selected amplitude and period values of pitch and roll movements, or regular, the mean of these values, and represented approximately 33% of the recommended maximum tolerance for livestock carriers. Pushing ($P=0.006$) and stepping behaviours ($P<0.05$) were increased when the stocking density was high but only when the sheep experienced regular motion, the latter particularly at the beginning of the treatment ($P\leq0.001$). The high stocking density and regular motion also decreased the RMSSD ($\sqrt{}$mean sums of squares of successive interbeat intervals, $P<0.0001$), indicating increased stress, and decreased time spent lying ($P\leq0.001$). Regular motion increased head lowered behaviour ($P=0.03$). Aggression (head to body attack) was increased by the high stocking density ($P=0.005$) and by irregular movement ($P=0.003$), the latter particularly at the end of the treatment ($P=0.05$). The ratio of low to high frequency beat intervals (LF/HF ratio) was reduced in the high stocking density, irregular movement treatment (($P=0.008$). Both motion treatments decreased rumination ($P=0.03$) and tended to increase affiliative behaviour ($P=0.06$) compared with control. Results suggest that there are two stresses caused by simulated ship motion. In the early stages of exposure sheep push each other in competition for space, especially at high stocking densities and when motion is regular, i.e. they can predict the position of others. In the later stages of exposure frustration builds in irregular movement and high stocking density and aggression results. It is concluded that a high stocking density increases both competition for space and aggression in sheep.

Effects of acute lying and sleep deprivation on behaviour and milk production of lactating Holstein dairy cows

Jessie A. Kull[1], Gina M. Pighetti[1], Katy L. Proudfoot[2], Jeffrey M. Bewley[3], Bruce F. O'Hara[3], Kevin D. Donohue[3] and Peter D. Krawczel[1]

[1]*The University of Tennessee, 2506 River Drive, Knoxville, TN 37996, USA,* [2]*The Ohio State University, 1900 Coffey Road, Columbus, OH 43210, USA,* [3]*Unviesity of Kentucky, 900 W.P. Garrigus Building, Lexington, KY 40546, USA; jkull@vols.utk.edu*

Sufficient lying time is important for dairy cows, as cows are highly motivated to lie down after a period of deprivation. Lying deprivation can result in sleep loss, as cows acquire most of their sleep while lying down. Evidence from human and rodent research shows that sleep deprivation can impact immunity, metabolism, and performance, but little research has assessed the impact of sleep in dairy cows. The objective of this proof-of-concept study was to determine the impact of sleep deprivation with and without lying deprivation on milk production and lying behaviour of dairy cows. To ensure cows were not experiencing undo distress, they were screened every 30 minutes for a set of removal criteria established before the experiment. All procedures were approved by the University of Tennessee Institutional Animal Care and Use Committee. Data were collected from 12 mid-lactation Holstein cows (DIM=199±44 (mean ± SD)). Using a cross-over design, cows experienced 24 h of: (1) sleep deprivation without lying deprivation achieved by noise or physical contact when the cow's posture suggested the onset of sleep, and (2) lying deprivation created using by a wooden grid on the pen floor. EEGs were used to ensure cows experienced sleep loss. Treatments were separated by a 12-d washout period. Cows were housed in individual box stalls (mattress base with no bedding) for habituation (d -3 and -2), baseline (d -1), and treatment (d 0) days. After treatment, cows returned to a sand-bedded freestall pen for 7 d (d 1 to 7). Lying time and lying bout duration were recorded from d -1 to 7 using accelerometers. Milk production was recorded twice daily on the day after treatment (d 1). Data were analyzed using a mixed model in SAS including fixed effects of treatment (sleep or lying deprivation), day (-1 to 7) and a random effect of cow. When there was a significant treatment by day interaction ($P<0.05$), pair-wise comparisons were made using the PDIFF statement. There was no effect of treatment on lying time or lying bout duration, but there was an interaction between treatment and day for both variables ($P<0.001$). As expected, lying time was lower when the cows were deprived of lying compared to sleep on d 0 (1.9 vs 8.4±0.7 h/d; $P<0.001$), but this difference was reversed on d 1 (16.8 vs 13.6±0.7 h/d; $P=0.002$). Similarly, lying bout duration was lower during lying deprivation than sleep deprivation on d 0 (15.3 vs 72.9±7.5 min/bout; $P<0.001$), and was reversed on d 1 (110 vs 89.9±6.8 min/bout; $P=0.01$). On d 1, milk yield was lower when cows were deprived of lying compared to sleep (31.8±vs 35.3±2.4 kg/d; $P=0.002$). The results indicate that 24 h of lying deprivation resulted in a rebound effect of lying time the day after deprivation. Lying deprivation also reduced milk yield compared to sleep deprivation, but more research is needed to determine if sleep deprivation alone has other negative impacts on the cow such as poor metabolic health and immunity.

Can faecal cortisol metabolites be used as an Iceberg indicator in the on-farm welfare assessment system WelFur-Mink?

Anna Feldberg Marsbøll[1], Steen Henrik Møller[1], Tine Rousing[1], Torben Larsen[1], Rupert Palme[2] and Jens Malmkvist[1]
[1]*Aarhus University, Department of Animal Science, Blichers Allé 20, 8830 Tjele, Denmark,* [2]*University of Veterinary Medicine, Department for Biomedical Sciences, Veterinärplatz 1, 1210 Vienna, Austria; anna.marsboll@anis.au.dk*

Faecal cortisol metabolites (FCM) are a non-invasive measure of HPA-axis activity. FCM have been found to relate to several aspects of welfare in mink, e.g. FCM levels was reduced by an enriched housing environment, and stereotypic mink had higher FCM levels that non-stereotypic mink. An on-farm welfare assessment system for mink named WelFur-Mink has been developed based on the concept of Welfare Quality*. In this system, a variety of animal- and resource-based measurements are taken on a sample of mink and aggregated into an overall welfare assessment at farm level. As FCM level is linked to stress responses in mink, it is expected to sum up several measurements in the system. We, therefore, aim to investigate if FCM can be used as an 'Iceberg Indicator', i.e. a welfare indicator that summarises many measurements and provides an overall assessment of welfare, with the potential to be used as a more cost-beneficial measurement in WelFur-Mink. In February 2016, the study was initiated with data collection on two Danish mink farms. 150 brown 1st years female mink were included on each farm. Faecal samples were collected 0-5 hours after feeding and the welfare of the individual mink was assessed according to the WelFur-Mink protocol. This includes assessment of stereotypic behaviour, response in a voluntary approach-avoidance test, fur chewing, body condition, and access to bedding material and enrichments. The behavioural measurements were taken first to avoid that the response was affected by the assessors walking behind the cages when collecting the faecal samples. The remaining measurements were taken after collection of faecal samples. Faecal samples were collected from 146 mink on farm 1 and 143 mink on farm 2. FCM levels (ng/g), indicative of circulating cortisol as validated for mink, were analysed using linear models, and differed between the two farms (median [2.5%; 97.5%]: farm 1: 277 [193; 398]; farm 2: 427 [238; 769], F=14, DF=1, P<0.001, ANOVA). The two farms also differed in the prevalence of fur chewing (Farm 1: 32%; Farm 2: 15%, P<0.001, Fisher's exact test), nest boxes with reduced thermal protection (Farm 1: 0%; Farm 2: 48%, P<0.001, Fisher's exact test), and stereotypic behaviour (Farm 1: 44%; Farm 2: 9%, P<0.001, Fisher's exact test). However, it was not possible to standardise the assessment of stereotypic behaviour according to feeding time, thus these prevalences are not comparable. The initial results indicate that FCM may be used to distinguish between farms. In February 2017, data collection on additional five farms is carried out based on the practical experiences from 2016. It will be investigated if mink with a reduced welfare according to WelFur-Mink have higher FCM levels, and if FCM can be used to identify farms with a reduced animal welfare according to WelFur-Mink. This will be investigated in relation to both individual and groups of welfare measurements, as well as the aggregated welfare assessment at criteria, principal, and overall assessment level.

Key-note:
The effect of a low dose of lipopolysaccharide on physiology and behaviour of individually housed pigs

Janicke Nordgreen[1], Camilla Munsterhjelm[2], Frida Aae[1], Anastasija Popova[1], Preben Boysen[3], Birgit Ranheim[1], Mari Heinonen[2], Joanna Raszplewicz[4], Petteri Piepponen[5], Andreas Lervik[6], Anna Valros[2] and Andrew M. Janczak[1]

[1]Faculty of Veterinary Medicine/Norwegian University of Life Sciences, Production Animal Clinical Science, Ullevålsveien 72, 0033 Oslo, Norway, [2]University of Helsinki, Production Animal Medicine, P.O. Box 3, 00014 Helsinki, Finland, [3]Faculty of Veterinary Medicine/Norwegian University of Life Sciences, Food Safety and Infection Biology, Ullevålsveien 72, 0033 Oslo, Norway, [4]Small Animal Teaching Hospital/University of Liverpool, Chester High Road, Neston CH64 7TE, United Kingdom, [5]Faculty of Pharmacy/University of Helsinki, Division of Pharmacology and Pharmacotherapy, P.O. Box 56, 00014 Helsinki, Finland, [6]Faculty of Veterinary Medicine/ Norwegian University of Life Sciences, Companion Animal Clinical Science, Ullevålsveien 72, 0033 Oslo, Norway; janicke.nordgreen@nmbu.no

Sickness may bring about changes in behaviour that disrupt social interactions. Production animals cannot withdraw socially while sickness lasts. Bad health is a risk factor for tail biting in fattening pigs. One possible mechanism is through cytokine effects on social behaviour, such as during underlying infection when bacterial products like lipopolysaccharide (LPS) may reach the circulation. Before this hypothesis can be tested, detailed knowledge of the changes in behaviour, mood and physiology after immune activation is needed. We describe the changes over three days in time budget, anticipation, cytokines, C-reactive protein, cortisol and immune cells in blood after injection with a low dose of LPS in gilts with permanent central venous catheters in the internal jugular vein. Brain levels of cytokines and monoamines were measured after euthanasia at 72 hours after injection. Pigs were injected IV with an LPS (O111:B4) dose of 1.5 µg/kg (n=7) or a similar volume of saline (n=6). An increase in cortisol ($F_{10, 110}$=14.05; P<0.0001), TNFα ($F_{10, 110}$=50.89; P<0.0001), IL-1ra ($F_{10, 110}$=62.26; P<0.0001), IL-6 (P=0.0034) and IL-8 ($F_{10, 110}$=27.44; P<0.0001) was seen within the first 6 hours after injection. During that period, activity decreased and sleeping increased in LPS pigs (P<0.05). Pigs were housed in pairs with visual contact (one saline, one control). The level of synchronization did not change significantly during the experiment. No effect on anticipatory behaviour was found. CRP was elevated at 12 and 24 hours after injection (P<0.001 for both) and food intake was lower for the first 24 hours (F 3,33=4,03; P=0.01). Absolute count of peripheral blood mononuclear cells was increased at 48 and 72 h p.i. Three days after the injection the pigs were euthanized by administration of an anaesthetic mixture followed by pentobarbital through the jugular catheter. LPS pigs had lower levels of noradrenaline in their hypothalamus (P<0.062), hippocampus (P<0.042) and frontal cortex (P<0.0006) compared to saline pigs. Thus, a low dose of LPS can induce changes in neurotransmitter levels that persist after inflammatory and stress markers in the periphery have returned to baseline levels. Future field studies on the effects of sickness on behaviour would benefit from following the animals until the clinical symptoms have resolved.

Effects of embryonic norepinephrine on juvenile and mature quail behaviors

Jasmine Mengers and Rachel Dennis
University of Maryland, Animal and Avian Science, 8127 Regents Drive, College Park, MD 20742,
USA; jmengers@gmail.com

Poultry breeding flocks experience stress from numerous sources including feed restrictions, confinement, social aggression, changing environments, transport, and stockperson turnover. Maternal diet and stress can increase catecholamine levels, including norepinephrine (NE), and alter tyrosine metabolism. Elevated NE levels impact the embryo and lead to altered survival related behaviors in the developed offspring, such as eating, drinking, and vigilance. In order to determine the effects of NE on feeding and social behaviors, activity level, and fear response, Japanese quail (*Coturnix japonica*) embryos were injected with 10 µl of 0.01 M or 0.05 M of NE or saline at ED1 (n=130) and incubated with intact controls (n=80). Weekly behavioral scan samples (am and pm) were taken from wk 4 (juvenile) to wk 11 (sexually mature) with rehoming between wk 6 and 7 samplings. Body weights were taken every other wk and organ weights were taken at wk 11. Tonic Immobility (TI) tests were conducted at 2, 5, and 9 wks. Data were analyzed in SPSS using ANOVAs with Bonferroni post hoc testing between treatments. Results showed a greater incidence of eating behavior in birds that received 0.01 M and 0.05 M of NE compared to control birds, with eating frequency increasing over time in older NE-treated birds ($P<0.05$ wks 5, 10, 11). Younger birds that received 0.01 M of NE spent more time drinking compared to birds that received 0.05 M of NE or controls ($P<0.05$ wks 5, 6). Birds that received 0.05 M of NE foraged more frequently compared to birds that received 0.01 M of NE ($P<0.05$ wks 6, 8, 10, 11). Following rehoming, NE-treated birds exhibited an increase in inactivity compared to control birds ($P=0.007$, wk 7). Birds that received 0.01 M of NE weighed significantly less than saline birds ($P\leq0.01$ wks 3, 5, 7, 9). Relative heart, liver, and spleen weights did not significantly differ between treatments. We observed no difference in TI inductions or duration. Our data show that NE injections during early embryonic development have altered feeding and drinking behaviors, activity levels, and weight as well as response to rehoming. These results suggest that increased adrenal hormone *in ovo* can impact both production and behavior, carrying implications for poultry management and indicating a need for further research into maternal stress in birds.

The influence of social stress on three selected lines of laying hens

Patrick Birkl[1], Joergen Kjaer[2], Peter McBride[1], Paul Forsythe[3] and Alexandra Harlander-Matauschek[1]
[1]*University of Guelph, 50 Stone Rd E, ON N1G 2W1 Guelph, Canada,* [2]*Friedrich-Löffler Institut, Dörnbergstraße 25, 29223 Celle, Germany,* [3]*Brain Body Institute, 1280 Main St W, ON L8S 4L8 Hamilton, Canada; pbirkl@uoguelph.ca*

When a social animal encounters a group of unfamiliar conspecifics, it faces the challenge of earning itself a spot in their well-established social web. For animals, and even humans, this can involve a range of interindividual interactions which can cause social stress. The aim of this study was to investigate the effects of social stress on three lines of laying hens (HFP, LFP and control). Feather pecking presents an abnormal behavior in laying hens. If this abnormal behavior was indicative of a compromised stress-response, then, we predicted that the HFP genotype would respond more sensitively to social stress in terms of weight-gain variability, changes in amino acid ratios that indicate a stress-response (phenylalanine: tyrosine, PHE: TYR), and increased occurrence of abnormal behavior (feather pecking). For this study, we used 160 laying hens selected for high (HFP) or low (LFP) feather-pecking activity and an unselected control line (C). Birds were housed in floor pens in groups of 16 hens per pen (HFP; n=4, LFP; n=3, C; n=9, 10 pens). At 16 weeks of age, we disrupted the groups in 5 pens by mixing individuals with unfamiliar birds to simulate social stress (this procedure was repeated after two days, to intensify social stress). Procedures were approved by the local animal care committee. Body weight was measured before mixing and again three weeks after the second mixing, to assess weight-gain. Blood plasma was collected one day prior to mixing and two days after the second round of mixing, to determine amino acid concentrations for calculating PHE/TYR ratios. Aggressive pecking and feather pecking were recorded on an all occurrence basis, prior to mixing (baseline), two minutes after mixing, one hour after mixing, and 24 hours after mixing, during 10-minute observation periods per pen and time-point. Data were analysed using a GLIMMIX procedure in SAS. We found that weight-gain showed increased variability for mixed groups, independent of genotype, compared to control groups (mixed: 15.1±0.9 vs unmixed: 4.9±0.9%, P<0.001). PHE/TYR ratios differed between all three lines (HFP: 0.75±0.01 vs LFP: 0.62±0.01 vs C: 0.65±0.01) and in HFP birds exclusively, PHE/TYR ratios were affected by social stress. Socially stressed HFP birds showed significantly lower PHE/TYR ratios than non-stressed birds (HFP unmixed: 0.75±0.02 vs HFP mixed: 0.70±0.02, P<0.02). Occurrence of aggressive pecking was increased two minutes after mixing (mixed: 4.5±0.9 vs unmixed: 1.0±0.5, P<0.05) but was not different from baseline levels one hour post-mixing and 24 hours post-mixing. Social stress had no short-term effect on the occurrence of feather pecking; yet ongoing analysis will investigate potential long term effects. The high variability of weight gain in mixed groups demonstrates physiological consequences of social stress beyond the increased occurrence of aggressive pecking. Our results also show that HFP birds differ in PHE/TYR ratios compared to other genotypes and that social stress reduces this ratio exclusively in HFP birds.

Feather pecking: is it in the way hens cope with stress?

Jerine A.J. Van Der Eijk[1,2], Aart Lammers[1] and T. Bas Rodenburg[1,2]
[1]Wageningen University, Adaptation Physiology Group, De Elst 1, 6708 WD Wageningen, the Netherlands, [2]Wageningen University, Behavioural Ecology Group, De Elst 1, 6708 WD Wageningen, the Netherlands; jerine.vandereijk@wur.nl

Feather pecking is a serious welfare and economic issue in the laying hen industry. It involves hens pecking and pulling at feathers or tissue of conspecifics, negatively affecting welfare. Excessive damaging behaviours, such as severe feather pecking, are indicative of an animal's inability to cope with the restrictive husbandry environment. The hypothalamic-pituitary-adrenal axis and the serotonergic system play an important role in how animals cope with stress and fear. Therefore, we here investigated whether lines divergently selected on feather pecking behaviour differ in their coping style, stress response and serotonergic system. We used genetic lines selected for high (HFP) and low (LFP) feather pecking and an unselected control line (CON). Lines were housed separately in groups of 19 birds per pen, with 8 pens per line. Group size was reduced by 2-3 birds at 0, 5, 10 and 20 weeks of age. There were two batches that differed 2 weeks in age. At 24 weeks of age birds were subjected to a 5 min. manual restraint test. We recorded latency to vocalize and struggle as well as number of vocalizations and struggles. No distincton was made between types of vocalizations. Blood was collected 10 min. after the test was finished for analysis of plasma-corticosterone levels and whole-blood serotonin levels. Corticosterone concentration was determined by a radio-immunoassay kit. Serotonin concentrations were assessed by a fluorescence assay. Data were analysed using mixed models, with selection line, age, batch, experimenter and test time as fixed factors and pen as random factor. HFP birds had a shorter latency to vocalize compared to CON and LFP birds (HFP=124 s, LFP=173 s and CON=185 s, $F_{2,180}$=5.77, P=0.0037). However, lines did not differ in their latency to struggle or in corticosterone levels after manual restraint. HFP birds did have lower serotonin levels compared to LFP and CON birds (HFP=31.98 nM/ml, LFP=35.84 nM/ml and CON=38.14 nM/ml, $F_{2,169}$=8.63, P=0.0003). These results indicate that HFP birds have a more pro-active coping style than CON and LFP birds. Previous behavioural tests at younger ages also indicated clear differences between the lines in coping style, with HFP birds having a more pro-active coping style in several behavioural tests compared to CON and LFP birds. In conclusion, our results suggest that selection for feather pecking might have an effect on coping style and this is accompanied by changes in the serotonergic system. However, selection did not have an effect on the stress response in this study. These results show that feather pecking seems to be related to coping style which might indicate that birds with a certain coping style are predisposed to develop feather pecking.

Do feather peckers, victims and control hens differ in symmetry of bilateral traits, body weight and comb size?

Fernanda M. Tahamtani[1], Lena K. Hinrichsen[1], Björn Forkman[2] and Anja B. Riber[1]
[1]Aarhus University, Department of Animal Science, Blichers Allé 20, Tjele, 8830, Denmark, [2]University of Copenhagen, Department of Veterinary and Animal Sciences, Grønnegårdsvej 8, 1870, Denmark; fernandatahamtani@anis.au.dk

Feather pecking is one of the major welfare issues facing the egg farming industry worldwide. Previous research has found a relationship between cannibalistic behaviour, fluctuating asymmetry of bilateral traits (FA) and body weight in laying hens. As cannibalism is linked to severe feather pecking, it could be suggested that a relationship between feather pecking, FA and body weight also exists. The purpose of this study was to analyse the association between feather pecking behaviour and (1) FA, (2) body weight and (3) comb size in laying hens. Sixty-four laying hens were categorised as feather peckers, victims or control hens based on weekly performance of severe feather pecking behaviour from age 0-23 weeks and their plumage condition at age 23 weeks. Hens with a high number of pecking bouts and low plumage damage score were classified as feather peckers. Hens with a low number of pecking bouts and high plumage damage score were classified as victims. Control hens had few pecking bouts and low plumage damage scores. After culling at 23 weeks of age, the lengths of ulna, length and width of tarsus, lengths of middle toe and widths of hock were measured twice in each side. Each trait was tested for repeatability, directional asymmetry and antisymmetry. Only traits that did not display directional asymmetry and/or antisymmetry were considered appropriate for analysis of FA. Composite FA was calculated using REML for the lengths of ulna, tarsus and middle toe. Control hens displayed less composite FA (0.43 ± 0.08 mm, P=0.0005) and less FA of ulna (0.10 ± 0.12 mm, P=0.0001) than feather peckers (composite: 0.83 ± 0.08 mm, ulna: 0.61 ± 0.11 mm) and victims (composite: 0.84 ± 0.08 mm, ulna: 0.69 ± 0.10 mm). Tarsus length asymmetry differed between all categories, with victims displaying most (0.82 ± 0.08 mm), control hens least (0.15 ± 0.09 mm) and feather peckers intermediate levels of asymmetry (0.40 ± 0.08 mm, P<0.0001). In addition, victims were also lighter in body weight ($1,728\pm43$ g) compared to control hens ($1,869\pm46$ g) and feather peckers ($1,859\pm45$ g, P=0.043). No difference was found in the size of the comb between the three categories (P=0.1). The results suggest that feather peckers and victims were exposed to similar levels of negative experiences, causing developmental instability, whereas control hens were less negatively affected during early life than both feather peckers and victims.

Investigation of response to the novel object test in pigs in relation to tail biting phenotypes

Jen-Yun Chou[1,2,3], Rick B. D'Eath[2], Dale Sandercock[2] and Keelin O'Driscoll[3]
[1]University of Edinburgh, Royal (Dick) School of Veterinary Studies, Easter Bush, EH25 9RG, Midlothian, United Kingdom, [2]SRUC, Animal & Veterinary Sciences Research Group, Roslin Building, Easter Bush, EH15 9RG, Midlothian, United Kingdom, [3]Teagasc, Pig Development Department, Moorepark, Fermoy, Co. Cork, Ireland; jenyun.chou@ed.ac.uk

Different physiological and behavioural traits of pigs engaged in tail biting have been widely studied, but most research has focused on identifying traits to predict tail biting phenotypes (biters, victims, neutrals). After pigs' involvement in tail biting, the short and long term characterisation of these traits is less understood. This study investigated if there was variation in stress and fear responses to a novel object (NO) test between pigs which were either performers or receivers of tail biting, or not involved in biting events, and whether these responses were still evident later in life. We hypothesised that 'biters' would exhibit a greater response than 'victims' or 'neutrals'. As part of a study on how different enrichment types (rubber floor toy or wooden post) and fibre content (high or low) in the diet affect the occurrence of tail biting, 72 undocked pigs (36 male/female) were selected for the NO test. The pigs were categorised as either a biter (n=24), a victim (n=24), or a neutral (n=24). The test was conducted twice for each pig: one week after they were moved into the finisher house (T1) when traces of biting were evident, and six weeks later (T2) when biting has already settled down. Pigs were placed individually in a test arena, and after a minute of habituation, a bright orange sweeping brush was introduced by dropping down from above. Direct continuous behaviour observation was conducted during the five-minute test. A saliva sample was taken immediately before (baseline) and after the test to determine salivary cortisol level. All data were analysed using mixed models (fixed effects: sex, category, treatment, T1/T2, before vs after; random effect: pen). There was no clear difference in the response to the NO test between the tail biting-associated phenotypes of biter, victim and neutral pig. Overall, cortisol concentration was higher after (0.234 ± 0.017 µg/dl) than before the test (0.152 ± 0.017 µg/dl; $P<0.001$), and the difference was much greater in T1 ($P<0.001$) than in T2 ($P=0.058$). Evidence of habituation to the test was also seen in behaviour: at T2 pigs had a lower frequency of vocalisation ($P<0.001$), longer latency to vocalise ($P=0.05$), longer duration of explorative behaviours ($P<0.001$), and approached the NO quicker ($P<0.001$) than at T1. Enrichment at the weaner stage had a significant effect on the latency to approach the NO ($P<0.05$); pigs given floor toys approached the NO faster than ones given wooden post (15.45 ± 5.53 vs 30.04 ± 5.46 s; $P<0.05$), and spent a longer time interacting with it ($P<0.05$), possibly because the NO was more similar to the floor toys and therefore was less novel and fearful for them. The habituation and enrichment effect both suggested the NO test was sensitive to detect different levels of stress and fear response, however, the preliminary evidence suggested there was no significant change in pigs' response after their involvement in tail biting events either in the short or long term.

Effects of elevated platforms on fearfulness in fast-growing broilers

Ida J. Pedersen[1], Fernanda M. Tahamtani[2] and Anja B. Riber[2]
[1]University of Copenhagen, Department of Veterinary and Animal Sciences, Grønnegårdsvej 8, 1870 Frederiksberg, Denmark, [2]Aarhus University, Department of Animal Science, Blichers Allé 20, 8830 Tjele, Denmark; idajp@sund.ku.dk

Fear is a general concern in the poultry industry. It may cause smothering, resulting in skin injuries and, in worst case, mortality. Thus, fear may result in detrimental animal welfare and economic loss to the farmer. Environmental complexity in the home environment of animals has been found to reduce fearfulness in several species, including mice, pigs and chickens. The aim of the present experiment was to investigate whether provision of elevated platforms with access ramps has an effect on fearfulness of fast-growing broilers, measured as their response to the tonic immobility test. Six pens (9.5×3 m) with approximately 500 broilers in each (Ross 308, mixed sex), corresponding to a stocking density of 40 kg/m2 were used. Commercial feed and water were provided *ad libitum*, and wood shavings were used as litter material. The birds were slaughtered at 35 days of age. The 3 treatment pens included an elevated platform (540×60×30 cm, length × width × height) made of perforated plastic slats with two access ramps. The area under the platforms was fenced off and was therefore not included in the net area used for calculation of stocking density. The 3 control pens did not include platforms, but otherwise the conditions were similar. At 31 and 32 days of age, 30 birds from each pen were randomly selected and induced into tonic immobility. The number of inductions needed for tonic immobility to be achieved (max. 3), the latency to first head movements (max. 10 min) and total duration of tonic immobility (max. 10 min) were registered for each bird. The statistical model was an ANOVA (GLM) with pen nested in block and treatment as a fixed factor. Birds with access to platforms had significantly shorter latencies to first head movement (149.5±135.2 s vs 230.0±191.7 s; $F_{1,175}$=9.84; P=0.002) and shorter total duration of tonic immobility (231.6±175.0 s vs 306.9±209.8 s; $F_{1,175}$=5.73; P=0.017) compared to control birds. No difference was found in the number of trials needed to induce TI between the control and treatment groups (1.15±0.42 vs 1.16±0.43; P=1.00). The results suggest that providing environmental enrichment in the form of elevated platforms can reduce fearfulness, and thus improve the welfare of fast-growing broilers.

Meloxicam and temperament effects on pain sensitivity and inflammatory response in surgical or rubber ring castrated calves

Désirée Soares[1], Sonia Marti[1], Daniela Melendez[1], Diego Moya[2], Eugene Janzen[1], Ed Pajor[1] and Karen Schwartzkopf-Genswein[3]
[1]University of Calgary, 3330 University Dr. NW, T2N 1N4 Calgary, Canada, [2]Aberystwyth University, Penglais Campus, SY23 3DD Aberystwyth, United Kingdom, [3]Agriculture and Agri-Food Canada, 5403 1 Ave S, T1J 4B1 Lethbridge, Canada; soares.desiree@hotmail.com

The majority of beef operations in Canada use surgical and rubber ring castration before 3 mo. of age and 90% do not use pain mitigation. Meloxicam is a nonsteroidal anti-inflammatory drug which has a label claim to control pain for up to 48 h. The role of animal temperament modulating acute and chronic pain caused by surgical and rubber ring castration is not known. The objectives were to determine the effect of meloxicam and temperament on pain sensitivity and inflammatory response in surgical and rubber ring castrated beef calves. Seventy-two Angus calves (76±2 d of age and 134.5±20.30 kg BW) were randomly assigned to treatments according to a 3×2 factorial design assessing castration technique (CAST): surgical (S), rubber ring (R) or sham castration as control (C) and drug administration (DRUG): single s.c. injection of meloxicam (M) at the time of castration (0.5 mg Metacam®/kg BW) or single s.c. injection of saline solution as control (N) to yield SM, SN, RM, RN, CM, and CN treatments (n=12/treatment). Calves were managed in two groups (GROUP) of 36 to be castrated on two separate days. They were housed on pasture with their mothers and with *ad libitum* access to water. Blood samples were collected via jugular venipuncture to assess plasma haptoglobin concentration (HP, g/l) as an indicator of inflammation. Pain sensitivity of the wound and surrounding skin was measured using a Von Frey anesthesiometer (VA, g). Temperament was assessed by using flight speed (FS, m/s). The variables HP, VA and FS were collected on d -1 and d 0 (immediately prior to castration) as baseline measures; and on d 6, 13, 20, 34, 48 and 62 post castration (SAMPLING DAY). Calves were blocked by the average FS and BW obtained on d -6 and d -1 (prior to castration day). Data was analyzed using a mixed-effects model including CAST, DRUG, SAMPLING DAY, GROUP and their interactions as fixed effects. The average baseline measurements of BW and each dependent variable as well as the average of all FS measurements were used as a covariate. A post-hoc (Tukey) test was used to compare the adjusted means. Our results showed that CAST, DRUG and FS had no effect on HP. Pain sensitivity was lower (P<0.05; greater VA values) in C compared to S and R calves. DRUG had no effect on VA. However, FS was related to pain sensitivity indicating that for each 1 m/s in additional FS the VA increased by 22.1 g (P<0.05). We conclude that a single s.c. injection of meloxicam at the time of castration had no effect on reducing indicators of pain or inflammatory response while more excitable animals showed less pain sensitivity over the 62 d post-castration. Further research is needed to improve the knowledge about pain mitigation post-castration in support of developing new strategies in beef calves and the relationship between temperament and pain sensitivity.

Cat responses to handling: assessment of scruffing, clips, and full body restraint

Carly Moody[1], Georgia Mason[2], Cate Dewey[1] and Lee Niel[1]
[1]*University of Guelph, Population Medicine, Ontario Veterinary College, University of Guelph, 50 Stone Road E., Guelph, Ontario, N1G 2W1, Canada,* [2]*University of Guelph, Animal Biosciences, Ontario Agricultural College, University of Guelph, 50 Stone Road E., Guelph, Ontario, N1G 2W1, Canada; cmoody@uoguelph.ca*

Cats require regular examinations to ensure optimal health; however, inadequate restraint can increase fear and aggression, resulting in inadequate physical examinations, and poor diagnosis and treatment. There is a lack of science-based evidence to inform best practice for restraint, so the current study evaluated the responses of companion cats to three common restraint techniques: (1) full body (n=19; known negative control); (2) scruff (n=17); and (3) clips applied to the neck (n=16). Cats in each group were also passively handled for use as a baseline comparison (known neutral control). Cats were blocked for sex and age, then randomly allocated to treatment groups. Order of restraint was also counterbalanced within groups. During each 1-min restraint, a mock veterinary examination was performed to assess previously validated measures of negative cat responses to handling. Behavioural (vocalizations- meow, yowl, growl; side and back ear position) and physiological (respiratory rate, pupil dilation) responses were assessed. Cats were initially assessed as either friendly or unfriendly during interactions with a stranger, and this was included as a covariate in analyses. Treatments were compared using linear mixed models that included the effects of sex and age, and cat as a random effect. Data are presented as averages with 95% confidence intervals. Overall, cats showed elevated responses to both full body and clip restraint. In comparison to passive restraint, cats showed larger pupil dilation, a greater number of vocalizations per minute, and a greater odds of having a negative ear position when handled with full body (pupils: 0.45[0.42,0.49] vs 0.54[0.49,0.59], P=0.003; vocalizations: 0.115[0.052,0.253] vs 0.388[0.159,0.947], P=0.009; ear: 0.336[0.170,0.662]; P=0.002), and clip restraint (pupils: 0.45[0.42,0.49] vs 0.53[0.48,0.58], P=0.006; vocalizations: 0.115[0.052,0.253] vs 0.510[0.189,1.374], P=0.007; ear: 0.441[0.220,0.8840], P=0.02). Respiratory rate was also 1.17× faster during full body vs passive restraint (CI: 1.049,1.309; P=0.006), but no effect was seen with the other treatments. In contrast, during scruff restraint cats had reduced pupil dilation and number of vocalizations when compared to full body (pupils: 0.41[0.36-0.47] vs 0.54[0.49,0.59], P=0.002; vocalizations: 0.057[0.011-0.299] vs 0.388[0.159,0.947], P=0.04) and clip restraint (pupils: 0.41[0.36,0.47] vs 0.53[0.48,0.58], P=0.003; vocalization: 0.057[0.011,0.299] vs 0.510[0.189,1.374], P=0.02). However, the odds of having a negative ear position was greater when handled with the scruff restraint in comparison to passive handling (0.346[0.161,0.743]; P=0.008). The current results indicate that responses to clip restraint are comparable to full body restraint, suggesting that the use of clips is negative for cats. While cats showed some negative responses to scruffing in comparison to passive restraint, it was less than that observed for full body or clip restraint.

Evaluation of saliva, milk and hair cortisol as an indicator for increased HPA axis activity in lactating sows

Winfried Otten[1], Susen Heimbürge[1], Margret Tuchscherer[1], Armin Tuchscherer[2] and Ellen Kanitz[1]

[1]Leibniz Institute for Farm Animal Biology (FBN), Institute of Behavioural Physiology, Wilhelm-Stahl-Allee 2, 18196 Dummerstorf, Germany, [2]Leibniz Institute for Farm Animal Biology (FBN), Institute of Genetics and Biometry, Wilhelm-Stahl-Allee 2, 18196 Dummerstorf, Germany; otten@fbn-dummerstorf.de

Minimally invasive evaluation of stress over short- and long-term periods is of increasing interest in biomedical and animal welfare research. Thus, the aim of this study was to investigate saliva, milk and hair cortisol concentrations after repeated administrations of adrenocorticotropic hormone (ACTH) as acute or long-term indicators for increased activity of the hypothalamic-pituitary-adrenocortical (HPA) axis in lactating sows. Fifteen multiparous German Landrace sows received either i.m. injections of ACTH (Synacthen Depot, 100 IU per animal) on day 1 until day 13 post-partum once a day (n=6), twice a day (n=5) or were left untreated (n=4). This procedure has been previously shown to increase plasma cortisol concentrations for several hours after each administration. During the 2-week treatment period, saliva and milk samples were taken on seven days and maternal behavior was observed. On day 1 and 14 post-partum, respectively, two square sections (15×15 cm) were shaved close to the skin in the caudodorsal region. Hair regrowth in these sections was collected on day 28 and 42, and additional samples of origin hair were taken. All experimental procedures were ethically approved by the Landesamt für Landwirtschaft, Lebensmittelsicherheit und Fischerei Mecklenburg-Vorpommern (LALLF, 7221.3-1-028/15). Cortisol concentrations in saliva, milk and hair were analyzed using a commercially available enzyme immunoassay either directly of after extraction procedures. Data were analyzed using pairwise multiple comparisons of the least square means by Tukey Kramer tests. Administrations of ACTH did not affect maternal behavior (e.g. pre-laying behavior, nursing frequency, crushing), but increased the cortisol concentrations in saliva (P<0.01) and milk (P<0.001) compared to control animals during the treatment period. On day 14, basal levels were reached 12 or 24 h after the last ACTH administration. Twice daily administration of ACTH caused higher cortisol concentrations in saliva compared to ACTH once a day (13.0±1.2 vs 7.9±1.1 ng/ml; P<0.05), while in milk no significant differences in cortisol levels between the ACTH treatments were found. Cortisol concentrations were increased in origin hair samples taken on day 42 compared to day 28 (P<0.05) independent of treatment, indicating a higher HPA activity after weaning at day 28 compared to preceding periods. Administration of ACTH generally enhanced hair cortisol concentrations in origin hair (P<0.05) and in hair samples with regrowth after day 14 (P<0.05). Our results indicate that an enhanced HPA axis activity in lactating sows is not only reflected in saliva, but also in milk cortisol concentrations. Hair cortisol can be a suitable and retrospective indicator for elevated glucocorticoids over longer periods in pigs and may be used for stress monitoring and severity assessment.

Effect on hair cortisol by the dominance relationship in the cow group

Hideaki Hayashi[1], Mami Matsumoto[1] and Shigeru Morita[2]
[1]School of Veterinary Medicine, Rakuno Gakuen University, 582 Bunkyodai-Midorimachi, Ebetsu, Hokkaido, 069-8501, Japan, [2]College of Agriculture, Food and Environment Sciences, Rakuno Gakuen University, 582 Bunkyodai-Midorimachi, Ebetsu, Hokkaido, 069-8501, Japan; hhayashi@rakuno.ac.jp

Dairy cattle experience stress as a consequence of management and production and it is desirable to reduce this as much as possible. Cortisol is a key hormone in the Hypothalamic-Pituitary-Adrenal axis (HPA) reaction, and blood and salivary cortisol are usually used for an index of the stress. However, blood and salivary cortisol exhibit circadian rhythm, and animals experience stress during sampling. Hair cortisol is not affected by the circadian rhythm, and reflects blood cortisol over the long term such as several weeks. However, it is unknown how the physiologic stress evaluation using hair and the behavioral evaluation are related. Therefore in this study we aimed to investigate effect of the dominance relationship in the cow group on hair cortisol. Thirteen Holstein cows a free-stall barn with an automatic milking system were used in this study. Using the dominance relationships for each cow, a dominance value (DV) was calculated: DV_i = (number of cows subordinate to cow i) / (number of known dominance relationships of cow i). Cows were divided equally based on DV to three groups of dominant, moderate and subordinate. Activity and rumination time were measured by attaching a 3 dimensional motion sensors tag to the neck of the cow. For each plasma sample, whole blood was collected via jugular venipuncture. White and black hair samples of approximately 1 g each were collected from the shoulder and the hip by severing the hair roots using scissors. Cortisol in collected hair samples was extracted with methanol after washing with isopropanol and then assayed using an enzyme immunoassay kit. The Student's t-test were employed for comparison between groups after ANOVA as the F test. Values were considered to be statistically significant if their P value was <0.05. The plasma cortisol concentration between dominant group and subordinate group was not significantly different. Shoulder white hair (1.12±0.09) and hip black hair (1.62±0.15) cortisol concentration of subordinate group were significantly higher than that (0.57±0.10, 0.97±0.15) of dominant group (P<0.05). Significant positive correlation was shown between plasma and shoulder black hair, plasma and shoulder white hair (0.82, 0.72, P<0.05). However, a significant correlation was not shown between hair and rumination time, activity, milk yield. These results suggest that subordinate cows suffer chronic stress from dominant cows and this stress can be evaluated using hair cortisol.

Behavioural effects at colostrum feeding when using an oesophageal tube or a nipple bottle in dairy calves

Carlos E. Hernandez[1], Bengt-Ove Rustas[1], Charlotte Berg[2], Helena Röcklinsberg[2], Stefan Alenius[3], Kerstin Svennersten-Sjaunja[1] and Lena Lidfors[2]
[1]*Swedish University of Agricultural Sciences, Department of Animal Nutrition and Management, P.O. Box 7024, 75007 Uppsala, Sweden,* [2]*Swedish University of Agricultural Sciences, Department of Animal Environment and Health, P.O. Box 234, 53223 Skara, Sweden,* [3]*Swedish University of Agricultural Sciences, Department of Clinical Sciences, P.O. Box 7054, 75007 Uppsala, Sweden; carlos.hernandez@slu.se*

New born calves have insufficient immunity to fight against disease and must therefore acquire maternal antibodies by ingesting colostrum soon after birth (transfer of passive immunity (TPI)). One of the main challenges in achieving an adequate TPI in calves fed artificially is the amount of colostrum consumed by the calves and time taken to bottle feed them. Calves that do not consume enough colostrum soon after birth are at increased risk of disease and mortality. For these reasons, some dairy farms force feed colostrum, using an oesophageal tube (OT), to all new born calves. While OT feeding seems ideal for TPI, it is not without risks. OT feeding is an invasive procedure that requires trained personnel to restrain and intubate the calves, thus increasing the risk of stress and discomfort. For these reasons, the aim of this study is to compare the effects of bottle vs OT feeding on the behavioural response of new born dairy calves. At birth, calves were randomly allocated to receive their first colostrum meal via an OT (n=15) or a nipple-bottle (n=30). Calves were fed colostrum 4 h after birth (an amount equal to 8.5% of the calf's body weight). Number of hind leg movements, vocalisations, struggles (i.e. push, slip or laying down) during feeding were analysed from videos. Differences for time feeding were compared using t-test, colostrum intake (express as % of colostrum allowance) with Chi-squared test, leg movements and struggles with Wilcoxon test. Vocalisations could not be analysed due to low frequency of the behaviour. Data are mean ± SEM. All procedures approved by ethics committee. Preliminary data from n=6 OT and n=8 bottle fed calves (analysis of remaining videos is in progress) shows that feeding calves using an OT is faster than with a bottle (370.8±121.8 vs 778.0±397.5 s, P<0.05, respectively). Colostrum intake was similar in both treatments. However, 2 bottle fed calves voluntarily consumed less than 60% of their colostrum allowance (remaining calves consumed >93% of their allowance) and 1 OT calf could only be fed 90% of its allowance due to excessive struggling (all other OT calves were fed >98% of their allowance). More calves vocalised during feeding using an OT than a bottle (3 vs 1 calves, respectively). No difference in the number of leg movements or struggles were found. Using an OT saves time during the first feeding and ensures all calves receive enough colostrum. However, from the welfare point of view, using an OT to feed colostrum to all new born calves appears to be unjustified given that only 25% of calves fail to drink enough colostrum and that 50% of the calves fed with an OT show some sign of distress (i.e. vocalisations). Therefore, preliminary data suggests that the use of an oesophageal tube feeder should be restricted to calves that fail to drink enough colostrum from a bottle.

Motor lateralisation as welfare risk measure in dogs entering a rescue centre

Shanis Barnard, Deborah L. Wells and Peter G. Hepper
Queen's University Belfast, Animal Behaviour Centre, School of Psychology, Malone Road, BT7 1NN, United Kingdom; s.barnard@qub.ac.uk

Previous studies have largely reported that shelter dogs may suffer poor welfare especially in the first few days following entry to a kennel environment. Commonly used welfare indicators are often resource- and time-consuming. The identification of quick and easy-to-assess measures to identify those animals that may be at welfare risk when kennelled, would enable a prompt intervention aimed at improving well-being. It has been suggested that motor bias (the preferred use of one limb over the other), has the potential to be used as an indicator of emotional functioning and welfare risk. Our aim was to investigate if motor laterality, in this case paw preference, could be used as a predictive indicator of stress (measured using cortisol levels and behavioural observation) in a sample of 35 dogs entering a rescue shelter. Based on previous literature, we hypothesised that higher stress levels would be associated to a left paw directional bias or a weak lateralisation. Early morning urine samples were collected on the animals' first day after kennel admission to assess cortisol levels. We recorded the animals while undisturbed in their kennel during three bouts of 30 min. observations and analysed the behaviour (e.g. locomotion, posture, position in kennel, vocalisations) as duration and frequency of occurrence. Finally, we scored the preferred paw used by the dog during a food-retrieval task (Kong™ test). Analysis carried out on 12 dogs to date showed a positive correlation between strength of laterality and barking behaviour (i.e. dogs spent more of their time barking with increased strength of paw preference; Spearman's rho: $R=0.62$, $P=0.03$, tendency after Bonferroni correction). No significant correlations emerged between the animals' cortisol levels and laterality (direction and strength of paw bias). Finally, we found significant correlations between cortisol level and behaviour: dogs with higher cortisol levels spent less time moving (Spearman's rho: $R=-0.67$, $P=0.017$) and barking ($R=-0.58$, $P=0.047$) and more time laying down ($R=0.66$, $P=0.018$) than animals with lower cortisol levels. Quieter dogs seamed to cope less in the new environment and tended to be weakly lateralised. These results only partially support our hypothesis. Our preliminary analysis suggest that there could be a relationship between motor laterality and stress levels in dogs entering a kennel; however, a larger sample is needed to confirm our results.

Effect of isolation stress on feeding behaviour of caged laying hens using a high-speed digital camera

Shuichi Ito[1], Mari Iwahara[1], Yumi Nozaki[1], Yuko Suenaga[1], Ken-ichi Yayou[2] and Chinobu Okamoto[1]
[1]Tokai University, Department of Agriculture, Kawayo, Minamiaso, Aso-gun, Kumamoto-ken, 869-1404, Japan, [2]National Agriculture and Food Research Organization, Animal Waste Management and Environment Research Division Institute of Livestock and Grassland Scien, 2, Ikenodai, Tsukuba, 305-0901, Japan; shuichi_ito@agri.u-tokai.ac.jp

Psychological stress affects feed intake and foraging behaviour in laying hens. Foraging behaviour of caged laying hens can be classified as true feeding, in which pecking is accompanied by food intake, and food tampering, in which pecking is not accompanied by food intake. It is unknown whether psychological stress affects true feeding and food tampering. In this study, we investigated the effect of stressors on laying hens and classified the foraging patterns using a high-speed camera. We randomly selected five commercial laying hens (Boris Brown) that were placed in individual conventional cages as the experimental hens. The hens were housed individually and fed a custom-made pellet diet *ad libitum*. During the control period, these birds were housed in the same room, whereas for social isolation, each test bird was isolated in its own cage by removing peers. The quantity of feed consumed was measured every hour on observation days using electrodes. We categorised feeding behaviours in detail by analysing images captured by conventional high-speed cameras. The number of pecks during each feeding behaviour was divided by the total number of times pecking. Statistical analysis was carried out using a Paired t-test for each feeding behaviour. The feeding behaviours classified by high-speed camera video analysis included hens showing 'true-feeding', 'feeding-failure (dropped the feed)' and 'food-tampering'. No differences were observed in the proportions of true-feeding (t=0.6247, df=4, P=0.566) or food-tampering (t=2.093, df=4, P=0.104) between the isolation and control conditions. However, the proportion of hens that feeding-failure was significantly (t=2.979, df=4, P=0.040) greater under the isolated condition (27.45±8.45%) than under the control condition (20.36±6.80%). The quantity of feed consumed was not different between the conditions. We observed an increase in the feeding failure rate in hens held under isolation stress; however, feed intake rate was not affected. Classifying feeding behaviour of caged hens using a high-speed digital camera may be useful to evaluate their behavioural response to stress.

Effects of overstocking and restricting feed access on the behaviour of lactating Holstein dairy cows

Mac A. Campbell[1,2], Peter D. Krawczel[3], Heather M. Dann[2] and Rick J. Grant[2]

[1]The University of Vermont, Department of Animal Science, 102 Terrill, 570 Main Street, Burlington, VT 05405, USA, [2]1William H. Miner Agricultural Research Institute, 1034 Miner Farm Road, Chazy, NY 12921, USA, [3]The University of Tennessee, Department of Animal Science, 2506 River Drive, 258 Brehm Animal Science, Knoxville, TN 37996, USA; krawczel@utk.edu

The cumulative negative effects of overstocked dairy housing and feed restriction, two common stressors, may be greater than either imposed individually. The objective was to determine the interaction between stocking density (100 vs 142% occupancy of feeding and resting space) and restricted feed availability (0 h vs 5 h of feed restriction) on lying, feeding, and rumination behaviours. Multiparous (n=48) and primiparous (n=20) Holstein cows were assigned to 1 of 4 pens (n=17 cows/pen) balanced for parity (2.3 ± 1.1; mean $\pm$ SD), days in milk (121 ± 38), and milk yield (47 ± 8 kg/d). Treatments were assigned to pens in a 4×4 Latin square with 14-d periods. A 2×2 factorial arrangement was used to combine two levels of stocking density and feed restriction into 4 total treatments. Total mixed ration was delivered 1×/d (approximately 06:00) and restriction began 19 h later. Time spent lying, feeding, and ruminating were measured using 10-min scan sampling for 72-h from d 8-10 of each period. Feeding and ruminating were summarized as h/d and total min from 0-8 h post-feeding and total min from 17-24 h post-feeding. Pen intake was measured on d 8-14. Data were analyzed using a mixed model in JMP with pen as the experiment unit. Contrary to our hypothesis, there was no interaction of stocking density and feed restriction on behavioural responses, which suggests that there was not a cumulative effect of these potential stressors. Overstocking decreased daily lying time (13.1 ± 0.2 (mean $\pm$ SE) vs 12.6 ± 0.2 h/d; P=0.02) and total rumination in a freestall (84.7 ± 1.4 vs $79.1\pm1.4\%$; P<0.01). Restricting feed access decreased total feeding time (3.8 ± 0.1 vs 3.7 ± 0.1 h/d; P<0.01). Feeding time in the 8 h following feed delivery was increased by restricting access (96.5 vs 85 ± 3 min), but decreased in the last hours before feed delivery (40 ± 2 vs 60.5 ± 2 min). Intake (25.8 ± 0.3 kg/cow/d) was not affected by this behavioural change (P>0.10). Total rumination time (518 ± 9 min/d) was not affected by feed restriction (P=0.43), but was decreased following feed delivery (153.5 ± 3 vs 162.5 ± 3 min/8-h period) and increased during the later period (199.5 ± 5 vs 187 ± 5 min/ 8-h period). While behavioral changes indicative of a cumulative effect of treatments were not evident, a key behavioural need (lying time) was reduced and diurnal patterns (feeding and ruminating) were altered to maintain total chewing activity. This suggests that both factors result in detrimental effects when presented independently of one another. The lack of a cumulative effect may indicate that cows use different strategies to mitigate these factors. Future research should focus on factors such as overstocking and freestall management, or restricted feed availability and diet composition, which might result in compounding detrimental effects on the behaviour of lactating dairy cows.

The behaviour and heart rate variability of wool-biting sheep

Akane Sato, Ken-ichi Takeda and Chen-Yu Huang
Faculty of Agriculture, 8304 Minami-Minowa-mura, Kamiina-gun, Nagano-ken, 399-4598, Japan; chenyo0601@icloud.com

Sheep subjected to wool-biting lose wool and their skin is injured. However, we previously found that some sheep did not try to escape when their wool was being bitten, in which the biters simultaneously ingested quantities of wool that could cause serious health problems. In this study, we attempted to identify (1) the bitten sheep's responses and the relationship between their posture and their hierarchical position and (2) the sheep's heart rate variability (HRV) stress response while having their wool bitten. This study examined 15 Friesland ewes. Each ewe was fed 1 kg of roughage daily at 08:30 and 15:00 and 350 g of concentrate at 15:20. Their behaviour was observed from 16:00 to 18:00, and the posture (standing/lying) and response (tolerate/escape) of the bitten sheep were recorded. The hierarchy of ewes was determined using the feed-scrambling method. We also used a heart rate sensor to record the R-R interval and HRV of four ewes while resting (standing/lying), and being subjected to wool-biting, brushing, wool-pulling and poking with a pencil. The heart rates and the ratio of low frequency to high frequency (LF/HF) in each situation were compared. Over 20 hours, 155 wool-biting events were recorded and 83.9% of the ewes escaped this behaviour rather than tolerated it (P<0.01). The ewes preferred to target higher-ranked individuals while they were lying down. Furthermore, ewes that were lying down tended not to escape when subjected to wool-biting. However, the heart rate and LF/HF were also higher in the lying resting ewes that did not escape. Our results suggest that sheep were under stress despite tolerating their wool being bitten by other individuals.

Does behaviour in novelty tests predict tail biting in pigs?

Sabine Dippel[1], Martina Gerken[2], Christina Veit[1] and Marie Albers[1,2]
[1]*Friedrich-Loeffler-Institut, Dörnbergstr. 25/27, 29223 Celle, Germany,* [2]*Georg-August-Universität Göttingen, Albrecht Thaer Weg 3, 37075 Göttingen, Germany; sabine.dippel@fli.de*

Stress can cause tail biting in pigs, which in some cases might be due to detrimental effects of stress on the affective state of the pigs. As affective state can influence how animals react to novel stimuli, we investigated whether behaviour of pigs in tests can predict the onset of tail biting under practical conditions. Tests procedures used were simple in order to allow potential daily use by farmers. A total of 17 groups of 35 to 40 pigs were tested on up to 12 test days in their home pens. The first test day was five days before, the last 24 days after groups were moved from rearing to fattening housing (no mixing). Testing stopped after severe tail lesions occurred. In each group, a voluntary human approach test (VHAT), a novel object test (NOT) and a modified novel object test where the object was thrown into the pen (mNOT) were performed on the same day with alternating objects. Parameters recorded during tests were seconds until first CONTACT and percent of group within one pig length of person/object at 30 seconds after start (INTEREST). During mNOT, 'percent of group STARTLED when object hits ground' was additionally recorded. After tests, pigs were scored for tail lesions. All tasks were performed by the same person. Tail biting outbreaks were defined as ≥20% of animals with tail lesions, and for each test, associations between test parameters and the independent factors 'DAYS in relation to outbreak' and 'consecutive test number' (Tno) were analysed in linear mixed models with batch (n=3) and group as random factors. No significant (P≤0.05) association between DAYS and any test parameter was found. Tail biting outbreaks occurred in all groups at median 6 [Q25=6|Q75=10] days after behaviour tests started, and many pigs already had ear lesions at that point. Tno was significantly associated with CONTACT (-0.83 s, SE=0.18) during VHAT, as well as with CONTACT (-0.82 s, SE=0.24), INTEREST (1.57%, SE=0.60), and STARTLE (-6.91%, SE=1.78) during mNOT. Median Tno per group was 10 [9|12]. Median CONTACT times were 0 s [0|1], 1 s [1|2] and 2 s [1|3] for VHAT, NOT and mNOT, respectively. Median INTEREST was 33% [29|36] during VHAT, 28% [22|33] during NOT and 26% [20|33] during mNOT. Median STARTLE was 40% [20|70]. The almost simultaneous outbreak of tail biting in relation to start of behaviour tests on this farm probably influenced the results due to confounding with habituation. Furthermore, the initial presence of ear lesions might have had an influence. However, although no association between behaviour of pigs in novelty tests and tail biting outbreaks could be established, the variation of behaviour between groups and over time warrant further investigation.

The occurrence of tail and ear lesions in relation to enrichment and manipulative behavior in weaner piglets

Christina Veit[1], Angelika Grümpel[1,2], Joachim Krieter[2] and Sabine Dippel[1]
[1]Institute of Animal Welfare and Animal Husbandry, Friedrich-Loeffler-Institut, Dörnbergstraße 25/27, 29223 Celle, Germany, [2]Institute of Animal Breeding and Husbandry, Christian-Albrechts-University, Hermann-Rodewald-Str. 6, 24118 Kiel, Germany; christina.veit@fli.de

Tail biting is closely linked with foraging and explorative behavior. While manipulative behavior directed to other pigs has been shown to increase during tail biting outbreaks in experiments, directing manipulative behavior away from pigs contributes to tail biting prevention. We wanted to find out, whether enrichment characteristics could be linked to manipulative behavior and tail and ear lesions of weaned piglets during on-farm assessment. During the application of a tail biting intervention tool in 132 pens on 27 farms, one observer assessed the proportion of active animals per pen which manipulated enrichment (ENRICH%), pen surroundings (PEN%), or other pigs (PIGS%) using scan sampling. Additionally, prevalence of tail and ear lesions and type of enrichment (solid objects only vs loose enrichment with/ without objects; n=42 vs 90 pens) were assessed, and a subjective estimate of 'percent of pigs which can simultaneously access enrichment' made. Data were analyzed using Wilcoxon rank sum tests. Due to skewed distribution, lesion prevalences were collapsed into binary variables using a threshold of 25%. The proportion of pigs manipulating enrichment was higher in pens with loose enrichment (all organic material, e.g. straw; median [Q25|Q75] = 9.5 [2.0|16.0]) compared to solid objects (e.g. objects on a chain; 3.0 [0.0|7.0]; P<0.001). This coincided with higher access for loose enrichment than for objects (median = 40 vs 20%, respectively). The prevalence of tail lesions was 20.5% [10.0 | 37.5] and ear lesions 19.0% [10.0 | 30.0]. Pigs in pens with ≥25% tail lesions (n=59) manipulated enrichment significantly more often (8.0 [3.0 | 13.0]) than pigs in pens with <25% tail lesions (n=73; 1.0 [0.0 | 7.0]; P<0.001). Regarding PIGS% or PEN%, no significant differences were found between pens with <25% tail lesions (median PIGS% and PEN%=0.0) and ≥25% tail lesions (PIGS% and PEN%=3.0). There were no significant differences for ENRICH%, PIGS% or PEN% between pens with <25% ear lesions (0, 0 and 4, respectively) and ≥25% ear lesions (0, 3 and 5, respectively). It can be concluded that provision of loose (organic) enrichment occupies more pigs in a group. Therefore, this combination of attractive material with better simultaneous access might be more effective in tail biting prevention on farm. The higher proportions of pigs using enrichment found in pens with ≥25% tail lesions might be an indicator for increased stress due to biting activities.

Intramuscular injections induce conditioned place aversion in calves

Thomas Ede, Marina Von Keyserlingk and Daniel Weary
University of British Columbia, Animal Welfare Program, 2329 West Mall, Vancouver, BC V6T
1Z4, Canada; thomas.ede92@gmail.com

Injections by needle are widely used in the administration of medicines, vaccination and anaesthesia. Injections cause somatic pain in humans, but little is known about how aversive these and alternative administration methods are for animals. We assessed aversion to injections using a conditioned place aversion protocol. Holstein calves (n=15) were randomly assigned to one of four routes of administration for 0.5 ml of saline: intramuscular (0.9×40 mm needle, rump), intranasal (MAD 300, right nostril), subcutaneous (0.9×20 mm, rump skin) and a null control (no treatment). The apparatus consisted of two distinctly colored connected pens with a milk reward available on one side. Approximately 12 h after their last milk meal, calves were allowed to enter the apparatus and received the treatment while consuming 1 l of milk from a bottle (for the first 3 sessions; this reward was reduced to 500 ml, and then to 250 ml, and finally to 0 ml for the last 3 sessions). Over all sessions, we observed a difference between treatments in the latency to enter the pen containing the milk (mixed model with repeated measures, P=0.005). Specified contrasts showed no difference in latency for the subcutaneous (mean ± SE: 15.3±5.0 s) and intranasal (9.6±6.2 s) treatments, with both being similar to the null control (4.4±0.9 s), and both shorter than for intramuscular injections (52.0±11.4 s). Latency increased with lower milk rewards (mixed model with repeated measures, P<0.0001), but there was no interaction with treatment. These results indicate that intramuscular injections are aversive, and that subcutaneous and intranasal injections are a refinement.

Pain face intensity for quantification of pain in Bos indicus **bull calves**

Karina Bech Gleerup[1], Teresa Collins[2], Gabrielle C. Musk[2], Timothy H. Hyndman[2], Heidi S. Lehmann[2], Craig B. Johnson[3] and Michael Laurence[2]
[1]University of Copenhagen, Department of Veterinary Clinical Sciences, Højbakkegård Alle 5, 2630 Tåstrup, Denmark, [2]Murdoch University, College of Veterinary Medicine, 90 South Street, Western Australia, 6150, Australia, [3]Massey University, Institute of Veterinary, Animal and Biomedical Sciences, Private Bag 11 222, 4442, Palmerston North, New Zealand; kbg@sund.ku.dk

Bos indicus bulls are most often castrated without the use of analgesia. To determine the most efficacious analgesic protocol, feasible pain evaluation methods are needed. The aim of this study was to investigate if pain face evaluations, using the bovine (Bos taurus) pain face, was a feasible method. Forty-one B. indicus bulls underwent halothane-anaesthesia allocated to four groups: one non-surgical group (n=5), and three castration treatment groups: CL, intra-testicular and subcutaneous lidocaine (n=12), CM, subcutaneous meloxicam (n=12) and C, no analgesia (n=12). The animals were observed before anaesthesia (t0) and after anaesthesia (t1, t2, t3 and t24) by the same observer (blinded to treatment), trained to evaluate the bovine pain face (B. taurus). The face was evaluated as a whole, as described for B. taurus but an intensity score was added to the pain face evaluation, using a simple descriptive pain face intensity scale, where 0=no pain face, 1 = mild intensity, 2 = moderate intensity, 3 = severe intensity. All ratings were performed by visual inspection from the distance through binoculars. Thirteen missing ratings were evaluated (blinded) from videos retrospectively. All animals scored 0 (no pain face) before anaesthesia. The non-surgical group scored 0 at every time point. For the castrated animals, the risk of receiving a pain face intensity score 2 or 3 during the first 24 hours following surgery was, 77% for group C, 80% for group CL and 53% for group CM. The results of this study indicate that, a direct pain face evaluation with an intensity score, may be useful for quantifying pain in B. indicus cattle. In this study, the animals treated with pre-operative meloxicam (CM), had a lower risk of receiving a moderate-severe intensity pain face score during the first 24 hours, compared to the animals in the C and CL groups.

'Sounding the slaughterhouse': testing the 'soundness' of cow vocalisations as an indicator of cow welfare assessment

Eimear McLoughlin

University of Exeter, Sociology, Philosophy and Anthropology, Amory Building, Rennes Drive, Streatham Campus, EX4 4RJ, United Kingdom; e.mcloughlin@exeter.ac.uk

In an auditory ethnography of the slaughterhouse, I will study how sound is managed, understood and mediated by lairage staff. I will explore how sound, from the biological, like cow vocalisations to the technological, like machinery, impact on and influence the lairage workers. In their daily routines, various sounds and noises are listened to and attended to, whilst others are heard and ignored. Lairage workers, drawing on a breadth of experience, tacit knowledge and certified animal welfare training, define what is important and what is unimportant throughout the day. In this manner, sound can be understood as a mode of surveillance in the lairage that is managed and negotiated by workers. In a sounding of the slaughterhouse, I contend that the lairage workers employ specialised knowledge in daily practices which enables rapid and informed welfare assessments. Lairage work is characterised by efficient and timely processing of animals that are in an unfamiliar environment and are likely to experience heightened levels of stress thus, I will interrogate how the brevity of contact from the human's perspective and the unfamiliarity of the setting from the animals' perspective can be negotiated whilst achieving an accurate welfare assessment. Drawing on the innovative approach of Wemelsfelder's Qualitative Behaviour Approach (QBA) to animal welfare which emphasises the 'wholeness' of the individual animal by integrating developments in philosophy of mind with biological animal welfare scholarship, I will explore how lairage workers assess animals for overall health and wellbeing, with a view to illustrating how assessments are a combination of objective indicators of the animal body and subjective interpretations of lairage workers. In order to validate the tacit knowledge of the lairage staff, I will focus on how cow vocalisations are perceived by workers, as these are typically used as quantitative stress indicators in welfare assessments. The complexity and contextuality of cattle vocalisations is experienced by lairage workers. Therefore, unlike previous animal welfare studies where only the frequency of cow vocalisations is of importance, this study will reproduce and explore the context of the calls, as experienced by lairage staff thereby contributing to a more holistic understanding of the wellbeing of the animal. This research will be conducted in May and will contribute to the growing literature into how qualitative behavioural assessments are both scientifically rigorous but adequately represent the complexity of animal sentience, thereby improving animal welfare standards. This research will unearth the intersection between sound, worker knowledge and welfare assessments thereby contributing to a deeper understanding of the relationship between lairage workers and cattle, as well as furthering slaughterhouse research on worker wellbeing.

Effect of long or short pre-slaughter handling on meat quality and physiological response of broiler chickens

A.F. Soleimani, A.O. Bello and Z. Idrus
Institute of Tropical Agriculture and Food Security, Universiti Putra Malaysia, Selangor, Malaysia; abdoreza@upm.edu.my

Finishing broiler chickens are commonly subjected to a range of stressors including catching, handling, crating, and transportation. These changes in a chicken's microenvironment activate stress response pathways in a range between mild to severe and affect health and welfare and meat quality. This study aimed to test the physiological effects of pre-slaughter handling in broiler chicken and conformed to the ISAE Ethical Guidelines. 96 day-old male broiler chicks were assigned in groups of 4 to 24 battery cages. On d 42, handling treatment was carried out over a short or long distance (25 or 100 meters). Handling involved catching, carrying the bird along the empty house, and ended with blood sampling. One bird per cage was caught at random with both hands, held in an inverted position and carried for a short or long distance (two birds per hands at once) through the house until reaching a separate room. The birds were then placed on a table and blood samples were drawn from wing vein. The average times between initial capture and end of blood sampling were 1:20 min for short handling group and 3:30 min for long handling group. Blood sampling procedure less than 3 min is considered to have no influence on corticosterone response. The same treatment procedure was carried out with another 24 birds, without blood sampling. These birds were crated (12 birds/crate) and loaded onto an open truck and transported for 2 h between 8:00 to 10:00 am with average speed of 80 km/h. Immediately upon arrival, blood and breast samples were collected for determination of drip loss, cooking loss, color and shear force at 24 h postmortem and serum levels of serotonin (5-HT), corticosterone (CORT) and heat shock protein 70 (HSP70). Data were analyzed as one-way ANOVA using SAS software package. The results revealed that the meat from short handling birds had greater drip loss and shear force compared to long handling group (5.49±0.26% and 103.29±4.82 N vs 4.28±0.33% and 91.61±2.86 N, respectively). Meat color was not affected by handling distance (P>0.05). The short handling birds also consistently showed higher HSP70 levels than their long handling counterparts (before transportation: 2.22±0.19 vs 1.73±0.09 ng/ml; after transportation: 2.17±0.18 vs 1.54±0.14 ng/ml). The 5-HT concentration increased in long handling birds compare to short handling groups only after transportation (36.64±3.57 vs 29.88±2.42 ng/ml). Although long handling resulted in higher levels of CORT than short handling before transportation (1.64±0.25 vs 1.00±0.14 ng/ml), it showed an opposite effect after transportation (0.98±0.08 vs 1.21±0.13). The present study showed that long handling, as practiced here, did not compromise welfare and meat quality of chicken. Long handling may decrease the incidence of PSE (pale, soft, exudative) meat in birds under transportation stress.

Key-note:

From Pavlov to Bekhterev – and the pros and cons of positive and negative reinforcement

Janne Winther Christensen[1], Andrew McLean[2] and Jan Ladewig[3]
[1]Aarhus University, Animal Science, Blichers Allé 20, 8830 Tjele, Denmark, [2]Australian Equine Behaviour Centre, 730 Clonbinane Rd, Clonbinane, 3658 Victoria, Australia, [3]University of Copenhagen, Veterinary and Animal Sciences, Grønnegårdsvej 8, 1870 Frederiksberg, Denmark; jwc@anis.au.dk

Pavlov and Bekhterev were rivals in their study of conditioned reflexes. Instead of using food to produce saliva, Bekhterev used mild electrical stimulation to study motor reflexes. According to Bekhterev, one of Pavlov's major research flaws included using the saliva method. If the animal is not hungry, food may not elicit the desired response, acting as evidence of the method's unreliability. Similarly, food motivation may influence an animal's performance in food-rewarded learning tasks. Nevertheless, applied animal ethologists often use food as a positive reinforcer in learning tasks to investigate the cognitive abilities of farm animals, and to test effects of various treatments, such as environmental enrichment, on animal cognition. One should be mindful, however, that any treatment that affects food motivation may also influence performance in food-rewarded learning tasks. Negative reinforcement (NR) is used less frequently in learning tests for farm animals and has been regarded as an inevitably negative (i.e. 'bad') training method, inducing poor welfare. However, recent research showed that horses trained through NR were more optimistic in a judgement bias test, compared to horses trained using positive reinforcement. Learning tasks based on NR may help circumvent some of the challenges with food motivation in food-rewarded learning tasks. Moreover, learning performance in positively and negatively reinforced tasks have been found to show poor correlations suggesting that interpretation of results based only on a food-rewarded test may not reflect the animal's learning performance in other reinforcement regimes. In addition, NR learning may be easily overlooked in experiments involving unpleasant stimulation. For example, studies on nociceptive thresholds (pain sensitivity) frequently use repeated stimulation. Typically the stimulation is removed when the animal shows an avoidance response, e.g. a foot-lift response, which is therefore negatively reinforced through removal of the stimulation. When stimulated again the animal is likely to show the same response at a lower threshold. Results may be interpreted as increased pain sensitivity whereas this sensitivity may in fact interact with NR learning. An increased focus on NR learning, and the interaction between animal motivation and the different learning modalities is likely to enhance the interpretative value of applied animal behaviour research.

Development of an ethical virtual fencing system for sheep

Danila Marini[1], Dennis Muleman[2], Sue Belson[3], Bas Rodenburg[2] and Caroline Lee[3]
[1]University of New England, Armidale, 2350, Australia, [2]Wageningen University, De Elst 1,
6700 Wageningen, the Netherlands, [3]CSIRO, New England Highway, 2350, Armidale, Australia;
dmarini2@une.edu.au

Virtual fencing (VF) has potential application in livestock management where the use of conventional fencing may be impractical. To ensure that animal welfare is not compromised using a VF system, the animal must be able to associate an audio cue with the negative stimulus (a mild electric shock), then learn that altering its behaviour in response to the audio cue will allow it to avoid receiving the negative stimulus. There has been limited work conducted on training sheep to VF. The aims of this study were to (1) find an appropriate stimulus level that would prevent sheep from reaching an attractant and (2) test if sheep can learn to associate an audio warning cue with an electric stimulus by responding to the audio cue alone. This study used 60, 4-year-old, Merino × Suffolk crossbred ewes with an average body weight of 66 kg. Electric dog collars which were manually controlled by a GPS monitor (Garmin, Australia) and had a stimulus range from 1 (low) to 18 (high) were used to deliver the audio and shock cues to the sheep. To determine an appropriate electric shock level, 30 sheep were trained to approach a feed trough and eat a feed reward (pellets). Once trained, sheep received a stimulus as they approached the trough, beginning at level 1 and increasing until all sheep displayed the desired response (to avoid the feed trough). Level 1 and 2 did not elicit a response in the sheep. When set to level 3, 1 out of 11 sheep reached the feed trough after receiving 5 stimulus applications. On level 4 no sheep reached the trough, this was determined to be an appropriate stimulus level. For the associative learning test, a further 30 sheep were trained to approach a feed trough. During the test sheep were fitted with the Garmin collar and allowed to approach the trough, at a distance of 4 m from the trough, an audio cue was applied for 1 s. If the sheep stopped or changed direction, the audio cue ceased immediately and no electric stimulus was applied. If the sheep did not respond to the 1 s audio cue, it was immediately followed by the electric stimuli (level 4) for 1 s. Testing ended after 5 min or if an individual sheep received a maximum of 5 shocks. The test was repeated once per day for 5 days. All 30 sheep approached the trough on day 1 and this was reduced to 8 on day 2, 9 on day 3, 11 on day 4 and 7 on day 5. There was a decrease in the number of shocks that were given across the days, however, this was not correlated to audio warnings. It is apparent that sheep readily learnt not to approach a trough after receiving a negative stimulus, however, the audio cues were ineffective in preventing sheep from receiving a shock. Further research is required to determine whether sheep can be trained to associate audio cues with a negative stimulus for use in virtual fencing.

Motor self-regulation by goats (*Capra aegagrus hircus*) in a detour-reaching task
Jan Langbein
Leibniz Institute for Farm Animal Biology (FBN), Institute of Behavioural Physiology, Wilhelm-Stahl-Allee 2, 18196, Germany; langbein@fbn-dummerstorf.de

Motor self-regulation, the ability to inhibit a prepotent motor response, is an important aspect of a cluster of processes generating effective inhibition of behaviour. In primates, it has been suggested that high level of sociability correlates with distinct inhibitory skills and as a consequence with higher behavioural flexibility. To broaden the range of species for a comprehensive discussion of this phenomenon, I studied motor self-regulation in a highly social animal, the goat. I investigated the behaviour of female dwarf goats (n=20; age 15-22 month) in a detour-reaching task, the cylinder task. Goats were trained to approach an opaque cylinder (length 20 cm × diameter 17 cm) from which they could retrieve a reward. After six correct trials in a row, subjects advanced to the test. In the test, an identical transparent cylinder was substituted for the opaque cylinder. Baiting of the cylinder by a human in full view of the animals was counterbalanced for side (left/right). I measured the goats' approach behaviour over a total of 10 trials. A trial was coded as 'accurate' if the goat retrieved the reward directly and 'inaccurate' if the goat tried to reach the visible reward directly by bumping into the cylinder. Additionally, I recorded duration of contact with the exterior of the cylinder, the side (left/right) the goat detoured the cylinder and latency to retrieve the reward. I used Generalised Linear Mixed Models to analyse the impact of trial on approach behaviour and of side of human baiting on the side the goats detoured the cylinder, while Mixed Linear Models were used to analyse the impact of trial on duration of contact and on latency to retrieve the reward. All statistical analyses were done using SAS. The overall average accuracy during testing was 62.5% (±SE: 10.86%). Trial had no impact on accuracy ($F_{9,171}=0.85$; n.s.). Latency to retrieve the reward decreased during testing (Max=35 s, Min=5 s; $F_{9,134}=7.85$; P<0.001), while duration of contact did not change (Mean=1.28 s; $F_{9,171}=0.93$; n.s.). The side of baiting the cylinder by a human had no impact on the side goats detoured it ($F_{1,151}=0.22$; n.s). For goats, living in complex social groups with a high level of fission-fusion dynamics, we found a level of behavioural inhibition equal or even higher compared to various bird and primate species. However, only a minority of subjects improved motor self-regulation during testing. This was in concordance with the constant duration of contact with the exterior of the cylinder across trials. Taken together, goats got faster in expressing the correct motor pattern to retrieve the reward but did not reliable improve inhibition of the prepotent motor response on the group level. Results show the importance to consider individual reaction patterns when analysing motor self-regulation. Effective behavioural inhibition might play an important role for group-housed farm animals living under restricted spatial conditions.

Cylinder size affects cat performance in the motor self-regulation task

Katarzyna Bobrowicz[1,2] and Mathias Osvath[2]
[1]*University of Warsaw, Faculty of Psychology, Stawki 5/7, 00-183 Warszawa, Poland,*
[2]*Lund University, Department of Philosophy, Helgonavägen 3, 22100 Lund, Sweden;*
katarzyna.bobrowicz@lucs.lu.se

We tested domestic cats in the so-called cylinder task, and found that they perform better if the cylinder is larger. We also found that their highest performance parallels that of great apes and corvids, which are known as the best performing animals on this task. The cylinder task is used to test animals' motor self-regulation: the inhibition of unproductive, but prepotent, movements in favour of productive movements that require a slight detour. Recently a large-scale study tested 36 species on this task and found that absolute brain size correlate with the performance; with great apes as top performers. Another study showed that corvids perform as good as great apes despite having smaller absolute brain size. We questioned whether average brained animals has as poor motor self-regulation as suggested, as it appears highly maladaptive; instead the results could be a reflection of the sensorimotor set-up of different species in relation to the materials used. No cats has yet been tested on the task. As ambush and sneak hunters, cats would arguably have high levels of motor self-regulation, but on the other hand their brain size and neuronal numbers are not above average in mammals. Eight adult domestic cats were tested in four versions of the task. We manipulated the size and materials to test whether that influenced performance: two large cylinders (16 cm diameter) out of glass and plastic respectively, and two small cylinders (9 cm diameter) of the same two materials. Each of the four conditions had two phases with a 24-hour delay in between. Each phase consisted of 10 consecutive trials. On the first day, a subject learned to retrieve a reward from an opaque cylinder. Next day, the cat was tested on a transparent cylinder. A retrieval of the reward without touching the cylinder's front counted as a successful trial. The success rate differed between conditions, and reached 98.75% in the 'big glass' condition, and 97.5% in the 'big plastic' condition, and 83.75% in the 'small glass', and finally 73.75% in the 'small plastic' condition. Two-Factor ANOVA for two within variables revealed a significant main effect of the cylinder size on the success rate [$F(1,7)=64.06$, $P<0.001$]. Neither a main effect of the material nor an interaction effect of size and material was statistically significant. The size effect was seen in all subjects. Failure rates did not decrease over time in any condition, so no learning curve was detected. Our results show that cats parallel great apes and corvids in the cylinder task as long as it is 16 cm in diameter and made of glass, despite their average mammalian neural characteristics. There are several possible explanations such as that a bigger size allows for more options of retrieval (e.g. mouth or paw), and/or requires less precise retrieval; it could also be that the distance to the reward is perceived as different. This calls into question whether the large-scale study took into account the sensorimotor architecture of each species, and more importantly, whether the task always measures motor self-regulation.

Factors affecting dairy calves' abilities to learn to use colour cues to locate a source of milk

Alison Vaughan, Jeffrey Rushen and Anne Marie De Passille
University of British Columbia, Main Mall, Vancouver, BC, Canada; jeffrushen@gmail.com

With the rise in automation in dairy farms it is important to understand the way cattle learn to interact with automated equipment. Dairy calves can have difficulties learning to use automated feeders or in learning to use a new food source, for example during weaning. Using colour cues to signpost resources may be an effective way of helping calves learn to use these food sources. We examined some factors affecting calves' ability to learn to associate colour cues with a milk source by placing unweaned dairy calves (2-6 weeks of age) in a Y maze. In one of the arms, a teat supplied milk while the teat in the other arm was dry. Colour cues (solid red or yellow boards) identified the arms with and without the milk. In exp. 1, calves (n=21) learned to locate the milk reward faster when the colour cues and the milk was always in the same arm of the Y-maze (fixed location) than when the colour cue and milk changed arms (changing location) (mean ±SD number of sessions: fixed side = 2.27±0.47 and changing side = 9.83±3.91; P<0.001). However, when calves initially trained with a fixed location were subsequently tested with changing location, they did no better (P>0.10) than those that initially trained with the changing location, suggesting that they were overlooking the colour cues altogether. In exp. 2, calves (n=8) were trained to find milk in one arm of a Y maze (fixed location) and then tested in a reversed Y maze (configuration rotated by 180 degrees). Most (7/8) calves turned to the same side (left or right) rather than choosing the same arm of the maze (P<0.001) suggesting that they had learned a particular response (e.g. 'turn left') rather than the location of the milk in the maze. In exp. 3 we investigated whether prior exposure to the colour cue outside of the Y maze would aid calves to subsequently learn to use colour cues to choose the correct arm. Calves (n=19) were either pre-trained to associate one colour (solid red or yellow) with milk reward or were exposed to colour cues which did not predict presence or absence of reward prior (control calves). When re-tested in the Y-maze with a variable location, most (9/10) pre-trained calves learned but no control calves learned to choose the correct colour cue within the allotted number of sessions. The most common error made by both pre-trained and control calves was to choose the same side as their previous choice. Most (5/7) control but no pre-trained calves displayed a clear side preference. When training calves to associate a milk reward with a colour cue, the colour cue should not be presented in a fixed location as the calves may ignore the colour and learn only to turn in a particular direction. Pre-training calves to use colour cues is effective at improving their performance in a Y maze test and may reduce the development of a side preference. Improving understanding of how cattle learn cues may help us improve training protocols.

Goats prefer happy human faces

Christian Nawroth[1,2], Natalia Albuquerque[3,4], Carine Savalli[5], Marie-Sophie Single[6] and Alan G. McElligott[2]

[1]Leibniz Institute for Farm Animal Biology, Institute of Behavioural Physiology, Wilhelm-Stahl-Allee 2, 18196 Dummerstorf, Germany, [2]Queen Mary University of London, Biological and Experimental Psychology, Mile End Road, London E1 4NS, United Kingdom, [3]University of Lincoln, School of Life Sciences, Brayford Pool, Lincoln LN6 7TS, United Kingdom, [4]University of São Paulo, Department of Experimental Psychology, Butantã, São Paulo 03178-200, Brazil, [5]Federal University of Sao Paulo, Department of Public Policies and Collective Health, Vila Clementino, São Paulo 04021-001, Brazil, [6]Technical University of Munich, Physiology Weihenstephan, Alte Akademie 8, 85354 Freising, Germany; nawroth.christian@gmail.com

Facial expressions are highly informative for humans and other animals. However, little is known about whether livestock animals perceive human emotional facial cues. In particular, discriminating emotional expressions in heterospecifics is considered especially challenging because emotions are not necessarily expressed in similar ways across species. We investigated whether goats show preferences for human faces with different emotional valences. To ensure exploration in subsequent test trials, goats first received a brief training session in which they had to approach an experimenter (with a neutral facial expression) at the opposite side of a test arena to receive a food reward. After training, two images of an unfamiliar human face (male or female) with different emotional valences (happy or angry) were presented to the subjects, one on the left and one on the right side of the arena. Goats (n=20) were free to explore the arena and interact with the stimuli; each test trial lasted 30 s. Each goat received a total of four trials, counterbalanced for position of the emotional faces (left vs right) and gender. We analysed first interaction, rate of interaction and time spent interacting with each face, using a Generalized Estimated Equation model for binary data with a logit link function. Goats interacted first, more often and for longer with the images on the right side (first interaction: X^2=9.63; P=0.0019; rate: X^2=10.47; P=0.0012; duration: X^2=7.67; P=0.0056). Moreover, we found an interaction effect of side (left vs right) and preference (happy vs angry): the side of presentation of the positive face affected the preference of goats for all three parameters (first interaction: X^2=8.90; P=0.0029; rate: X^2=8.01; P=0.0047; duration: X^2=8.24; P=0.0041). For happy faces, goats interacted first, more often and for longer when they were positioned on the right side. No preference was found when the happy faces were placed on the left side. Face gender and goat sex had no effect on their preference for any parameter. We show that goats can discriminate human facial expressions with different emotional valences and prefer happy ones.

Responses in novelty tests are associated with grain intake and unrewarded visits to the milk feeder in dairy calves

Heather W. Neave, Joao H.C. Costa, Katrina Rosenberger, Daniel M. Weary and Marina A.G. Von Keyserlingk
University of British Columbia, Animal Welfare Program, 2357 Main Mall, Vancouver, BC, V6T 1Z4, Canada; hwneave@gmail.com

The feeding behaviour of dairy calves is known to be highly variable between individuals. We tested the hypothesis that personality features, as revealed by standardized tests, can account for this variation. Our hypothesis was that more interactive and less fearful calves in novel situations would explore their feeding environment more than other individuals, particularly when calves are learning to eat solid feed or when they must find novel food sources when milk is limited. We also expected these calves to be more persistent in attempting to gain milk during weaning. Fifty-six dairy calves were housed in 7 groups, and randomly assigned within group to either 6, 8, 10 or 12 l/d of milk (n=14 each) until d 41 of age, when milk was reduced to 50% of allowance. At d 50, milk was gradually reduced by 20%/d until calves were completely weaned at d 55. Calves had *ad libitum* access to calf grain and hay. We measured DM intake of grain and the number of unrewarded visits to the robotic milk feeder during the 4 d before the 50% reduction of milk allowance (d 38-41), the day before weaning (d 49), the day after weaning (d 50), the 2 weeks after weaning (d 55-69), as well as for the total experimental period (d 7-69). At d 30 and d 70 of age, each calf was subjected to a human approach test (10 min with an unknown stationary human), a novel object test (15 min with a black 110 l bucket) and a food neophobia test (30 min with two identical white 20 l buckets, one empty and the other filled with a novel food, a fresh cow total mixed ration). Seven behaviours (vocalizations, time spent touching, latency to touch human/object, looking at or disengaged with human/object, time spent playing, and intake of novel food) were recorded, and data were averaged across age and then across tests. These behaviours were examined in a Principal Components Analysis where 2 factors explained 66% of the variance (Factor 1 = 49% and Factor 2 = 18%). Factor 1 had high positive loadings on time spent in contact with the human/object and time spent playing, and a high negative loading on time spent disengaged with the human/object. Factor 2 had high positive loadings on the number of vocalizations and time spent disengaged with human/object, and high negative loadings on time spent looking at human/object, latency to touch human/object and intake of novel food. Factor 1 and Factor 2 loadings for each calf were regressed against feeding behaviour variables for each feeding period with milk allowance, birth weight, sex and group included as covariates in the model. Calves with high Factor 2 loadings had greater starter DM intake during the day before weaning (R^2=0.33; P<0.01) and a higher number of unrewarded visits to the milk feeder during the total experimental period (R^2=0.61; P<0.01). These results indicate that personality differences, as revealed in standardized tests, are predictive of grain intake before weaning and of continued attempts to receive milk when it is unavailable.

Light during incubation and noise around hatching affect cognitive bias in laying hens

T. Bas Rodenburg[1,2], Nathalie J.T. Scholten[2] and Elske N. De Haas[1,2]
[1]*Wageningen University & Research, Adaptation Physiology Group, P.O. Box 338, 6700 AH Wageningen, the Netherlands,* [2]*Wageningen University & Research, Behavioural Ecology Group, P.O. Box 338, 6700 AH Wageningen, the Netherlands; bas.rodenburg@wur.nl*

Incubation and hatching conditions can influence development of laying hens. In commercial conditions, chicks are incubated in the dark and hatch in a noisy environment. This may affect their behavioural and cognitive development in a negative way and may have lasting effects on their assessment of ambiguous stimuli in a cognitive bias test. The aim of this study was to investigate the effects of light during incubation and noise around hatching on cognitive bias in adult laying hens. To meet that aim, eggs were incubated and hatched according to four different treatments in a 2×2 cross design: during the first 16 days of incubation, they would either be incubated in the dark or in a 12L:12D light schedule. From 17 to 21 days of age, half of the eggs from each light treatment would either be incubated in a noisy environment (playback commercial incubator noise; 90 dB) or a quiet environment (60-70 dB). Birds were kept in groups of maximum 10 birds in floor pens separated per treatment from 0 to 35 weeks of age (seven groups per treatment). At 35 weeks of age, 32 hens (eight per treatment) were trained for a cognitive bias test, where they had to learn that one side of the test arena was rewarded (mixed grains and mealworms) and the other side was not (smell of feed only). The rewarded side was varied according to a balanced design. The experimenter was blind to birds' treatments. When a bird had learned the task, she was tested with ambiguous cues, where the feeder would be placed in the centre (50%), nearer the unrewarded side (25%) or nearer the rewarded side (75%). This was done by subjecting each hen to 12 trials, where two rewarded (R) trials and one unrewarded (U) trial (order RRU or RUR, balanced design) were followed by one of the ambiguous trials. For all hens, the first ambiguous trial was the 50% trial, followed by the 25% trial and the 75% trial. Latency to approach the feeder was analysed using a mixed model with treatment, trial order and type of cue as fixed factors (including all interactions) and pen nested within treatment as random factor. Over all treatments, birds had a shorter latency to reach the feeder with the rewarded cue compared to all other cues, and with the 50% and 75% compared to the unrewarded and 25% cue as expected ($F4,360=211.4$; $P<0.001$). There was a significant interaction between treatment and type of cue ($F12,360=2.11$; $P<0.05$). Birds from the light and noise treatment were slower than birds from the other groups in the ambiguous 50% trial ($F1,360=3.49$; $P<0.05$), which is interpreted as a more pessimistic response to the ambiguous cue. The light treatment during incubation may have resulted in a faster embryo development. This may in turn have resulted in chicks that were more sensitive to the noise treatment that was applied around hatching. In conclusion, incubation and hatching conditions in laying hens can affect cognitive bias in laying hens. Creating more quiet incubators and could have positive effects on laying hen welfare.

Facial expressions of emotion influence sheep learning in a visual discrimination task

Lucille Bellegarde[1,2,3], Hans W. Erhard[2], Alexander Weiss[1], Alain Boissy[4] and Marie J. Haskell[3]
[1]The University of Edinburgh, School of Philosophy, Psychology and Language Sciences, EH89JZ Edinburgh, United Kingdom, [2]INRA, UMR Modélisation Systémique Appliquée aux Ruminants, Université Paris-Saclay, 75005 Paris, France, [3]SRUC, West Mains Road, EH93JG Edinburgh, United Kingdom, [4]INRA, UMR Herbivores, Centre de recherche Auvergne-Rhône-Alpes, 63122 Saint-Genes-Champanelle, France; lucille.bellegarde@sruc.ac.uk

Faces are an essential source of information for social species. Sheep are one of the most studied farm species in terms of their ability to process information from faces, but little is known about their face-based emotion recognition abilities. We investigated whether sheep could use images of faces displaying different emotional states as cues in a simultaneous discrimination task. To that end, we took photos of faces of four sheep in three social situations: two with a negative valence (social isolation or aggressive interaction) and a neutral situation (ruminating in the home pen). In a two-armed maze, sheep (n=16) were then presented with pairs of images of the same familiar individual taken in the neutral situation and one of the negative situations. Sheep had to learn to associate one type of image from a pair with a food reward. Once they had reached the learning criterion, sheep then had to generalise the task to new pairs of images of different conspecifics displaying the same emotions as in the previous phase. For every run in the maze, the latency to choose an arm and the outcome of the choice (success or error) were recorded, as well as the total number of runs needed to learn the task (learning speed). Influence of the type of image rewarded and of the side of presentation of the rewarded image were analysed by linear mixed models and learning speed was analysed by Mood's median test. All sheep learned the task with images of faces. Sheep that had to associate a negative image with a reward learned faster that sheep that had to learn the neutral-reward combination (medians: 40 vs 75 runs, χ^2=4.00, df=1, P=0.046). With the exception of sheep from the Aggression-rewarded group, sheep could generalise the discrimination task to images of new faces (F3,26.6=3.37, P=0.033). Sheep chose an arm correctly more often ($F_{1,288.1}$=5.02, P=0.026) and more quickly ($F_{1,289.3}$=23.92, P<0.001; right: 8.4±0.6 sec, left: 9.6±0.8 sec) when the rewarded image was displayed on the right side, suggesting the influence of a right hemisphere/left visual field bias in face-based perception of emotions. Our results strongly suggest that sheep can perceive the emotional valence displayed on faces and that this valence affects learning processes.

Free-choice exploration in laboratory-housed zebrafish

Courtney Graham, Marina A.G. Von Keyserlingk and Becca Franks
University of British Columbia, Applied Animal Biology, 2357 Main Mall, Vancouver BC, V6T
1Z4, Canada; courtney3graham@gmail.com

Cognitive stimulation has been shown to be rewarding and capable of eliciting positive emotions in several species. In contrast to the abundant learning and exploration opportunities available in nature, captive environments can be under-stimulating – with the potential to induce anhedonia and reduce welfare. Zebrafish are now a popular scientific model, in part because of their high cognitive function and sensitivity to environmental manipulations, yet little is known regarding their response to opportunities to explore. As a first step toward understanding the role that cognitive enrichment may play in zebrafish welfare, we housed zebrafish within six large structurally-enriched tanks (10 fish/110 l tank) for nine months and then removed a divider to expose 10 cm of additional novel tank space. This approach allowed us to measure free-choice exploratory behaviour (latency and number of inspections) as well as anxiety (bottom-dwelling) and social behaviour (aggression, cohesion and coordination). We collected video data on each of four days: the day before (baseline), the day of, the day after, and two weeks after divider removal. The number of inspections was scored in 30-sec bins (120 lines of data for each observation) and bottom-dwelling from spatial location was recorded every two minutes (32 lines) for one hour. Social behaviours were scored every two minutes from the time immediately after divider removal to tank lights out at 18:00 h (210 lines). Using linear mixed models, we found that after the divider was removed, zebrafish moved into the novel space on average within 9.7 ($\pm$7.6 SD) seconds, and that the number of inspections increased on each day observed ($P<0.003$). We found no evidence of bottom-dwelling ($P>0.73$), which, as a well-established measure of anxiety in zebrafish, indicates that the manipulation was not anxiety-provoking. Further, the opportunity to explore altered social behaviour: it reduced aggression on the day of and day after ($P=0.02$), and increased cohesion on the day of ($P=0.04$) and coordination on the day after ($P=0.04$) divider removal. These results contrast with previous research that has shown that zebrafish living in larger tanks tend to have lower group cohesion. Our finding that group cohesion *increased* in response to additional space suggests a different mechanism may be at play (i.e. cognitive rather than physical). In summary, we found that zebrafish readily engage in free-choice exploration and that such opportunities do not induce anxiety behaviour. Moreover, these opportunities alter social behaviour: reducing aggression and increasing cohesion and coordination. Considering that their natural environment would normally include many such information-gain opportunities, our findings suggest that high aggression and low cohesion and coordination may be signs of abnormal behaviour in captive zebrafish. This study provides novel evidence and adds to the growing body of literature on the role cognitive stimulation may play in zebrafish welfare. It indicates that zebrafish are good candidates for further cognitive enrichment research and we encourage future research to elucidate our findings.

Cognitive bias and paw preference in the domestic dog (*Canis familiaris*)

Deborah Wells, Peter Hepper, Adam Milligan and Shanis Barnard
Queen's University Belfast, Animal Behaviour Centre, School of Psychology, Queen's University
Belfast, BT7 1NN, United Kingdom; d.wells@qub.ac.uk

Cognitive bias measure, designed to detect the valance of affective states, have been proposed as a tool for assessing animal emotions and welfare risk. Such measures are not always practical in applied contexts. Limb use, an indicator of hemispheric functioning, may be a useful predictor of cognitive bias. This study explored the association between motor asymmetry and cognitive bias in the domestic dog. Thirty pet dogs had their paw preferences assessed using the Kong ball test, in which the animal had to use its paw/s to stabilise a food-filled ball. Affective state was assessed following training on a cognitive bias test used previously on dogs. Here, latency to approach a bowl placed in one of 3 ambiguous positions (referred to as Near Positive, Middle, Near Negative) between a baited food bowl (Positive) and a non-baited food bowl (Negative) was recorded. Positive bowls were placed, unblinded, on the left-hand side of the room for 50% of the dogs, and on the right side for the remaining subjects. Friedman tests were conducted for left, right and ambilateral animals to determine if their speed to approach the 3 ambiguous bowls differed according to location. Pearson moment correlations examined the relationship between the direction and strength of the dogs' paw preferences and their latency to reach the 3 ambiguous bowls. Kruskal-Wallis tests revealed no significant relationship between the dogs' paw preferences and their latency to approach the bowl located at the Negative (0.02, n=30, P=0.99) or Positive (1.57, n=30, P=0.46) position. Right-pawed (Friedman test=7.40,df=2, P=0.02) and ambilateral (Friedman test=22.15, df=2, P<0.001) dogs differed in their latency to approach the 3 ambiguous food bowls, becoming increasingly slower the further the bowl was positioned from the baited food bowl. Left-pawed animals showed no significant difference in their latency scores for the bowls positioned at the 3 ambiguous locations (Friedman test=3.71, df=2, P=0.16). Analysis revealed a positive correlation between the direction of the dogs' paw use scores and their latency to approach the food bowl positioned at Near Positive (r=0.66, n=30, P<0.001). Animas became slower to reach the bow located in this position with increasing left-pawedness. The strength of the dogs' paw preferences was correlated with their speed to reach the bowl located at Near Positive (r=0.39, n=30, P=0.03). Animas became slower to reach the bowl at this location with increasing lateralisation. The study points to a relationship between cognitive bias and paw preference in the dog, with left-pawed animals being more negative in their cognitive outlook than right-pawed or ambilateral individuals. Limb preference testing might offer a more practical way of identifying individuals at risk from poor welfare by virtue of how they perceive the world than more time-consuming cognitive bias tests.

Breed and behavioral characteristics of dogs that differ in owner reported tendency to attend to visual stimuli on a TV

Miki Kakinuma, Izuru Nose, Yumiko Uno, Shuichi Tsuchida, Hitoshi Hatakeyama, Sayaka Ogawa and Miyoko Matoba
Nippon Veterinary and LIfe Science University, 1-7-1 Kyonancho, Musashino-shi, Tokyo, 180-8642, Japan; kakinuma-miki@nvlu.ac.jp

Dogs are known to understand human forms of communication such as pointing and that they are born with the ability to communicate with humans. On the other hand, some researchers suggest that experience of living close to human is essential in doing so. Our previous study indicated that therapy dogs attend to visual stimuli on a TV screen more than regular pet dogs. We also found breed differences in attending to visual stimuli on a TV screen. Based on the description of infant TV watching behavior, such as attending to the screen, trying to touch the screen, talking to the person on the screen or looking behind the monitor, we asked owners for similar behavior in their pet dogs. We conducted the dog personality tests and collected oral swabs for DNA analysis and compared the results. In a study on dog behavior survey by owners 179 pet dog owners were recruited in Tokyo, Japan. We excluded 17 dogs that had no access to TV at home. Dog owners were asked to fill in a three page survey with dog breed and basic demographic information, reaction to a TV screen and dog personality test. Of the 161 dogs 31 were Toy or Miniature Poodles, 18 mixed breeds, 13 Shiba Inu, 11 Dachshunds and 10 Border Collies. Of those with access to TV, 52 attended to TV screens. Border Collie attended to the TV screens, Dachshund did not (χ^2 (3)=16.10, P<0.01). The result of dog personality test showed that TV screen attending dogs were higher in 'trainability' than non-attending dogs (t(161)=4.44, P<0.01). TV attending Poodles scored higher in 'trainability' than non-attending Poodles (t(31)=3.26, P<0.01) In a study on DNA analyses, oral swabs of 120 pet Toy Poodles were collected. The genomic DNA samples were extracted from oral swabs using DNA extraction kit. We focused on a gene, *FNBP1L*, which was estimated to be one of associated polygenic genes to human intelligence by gene-wide association study. The 14 coding regions in canine *FNBP1L* gene were amplified by PCR and carried out direct sequencing. By comparing the nucleotide sequences of the *FNBP1L* gene, the efficiency of the gene is evaluated in the dog behavior concerning attention to TV screens. The analysis of results is still underway. Since small mall breeds are popular in Japan, our findings may not apply to other societies with larger breeds or outdoor dogs. Our findings suggest that dogs attending to TV may have better relationship with the owners, thus are more adapted to the human society. The DNA analysis may help us understand the meaning of the owner reported personality traits. Nippon Veterinary and Life Science University ethics committee approved the experiments (S27K46). This research was supported by MEXT-KAKENHI Grant Number 15K04086.

Effects of birth weight and gender on spatial learning and memory in pigs

Sanne Roelofs[1,2], Ilse Van Bommel[2,3], Steffi Melis[2,4], Franz Josef Van Der Staay[1,2] and Rebecca Nordquist[1,2]

[1]*Brain Center Rudolf Magnus, Universiteitsweg 100, 3584 CG Utrecht, the Netherlands,* [2]*Utrecht University, Department of Farm Animal Health, Yalelaan 7, 3584 CL Utrecht, the Netherlands,* [3]*Utrecht University, Graduate School of Life Sciences, Heidelberglaan 8, 3584 CS Utrecht, the Netherlands,* [4]*HAS University of Applied Sciences, Onderwijsboulevard 221, 5223 DE 's Hertogenbosch, the Netherlands; s.roelofs@uu.nl*

In current pig farming, selection for large litter sizes has led to an increased number of low birth weight (LBW) piglets being born and surviving to slaughter age. In humans, LBW is a risk factor for various cognitive deficits. In pigs, the consequences of LBW are less well known. Cognitive deficits may influence animal welfare if they result in an impaired ability to cope with housing and rearing conditions. Previous studies have provided contradictory results when comparing cognitive performance of normal birth weight (NBW) and LBW piglets in a spatial holeboard task. It is possible that gender acted as a confounding factor on these results, as both males and females were tested without using gender as a factor in statistical analysis. Therefore, this study compared cognitive performance of LBW and NBW piglets of both sexes. Forty piglets from 15 available litters were selected (as NBW-LBW sibling pairs of the same gender). This resulted in twenty NBW piglets (1,446±49 g) and twenty LBW piglets (810±24 g), each group consisting of ten males and ten females. They were moved to the research facility at approximately 4 weeks of age and housed in groups of 10 in mixed sex groups, but separated by birth weight. All piglets were housed in pens enriched with straw. The piglets were trained to search for the locations of four hidden food rewards (chocolate M&M's) in the spatial holeboard, an open arena containing a 4×4 matrix of potentially baited holes. The holeboard allows for the simultaneous assessment of multiple cognitive variables, such as working memory (assessed by how many revisits a pig makes during a trial) and reference memory (assessed by how well a pig learns its rewarded configuration). All piglets received 44 acquisition trials. Testing in the holeboard lasted until the pigs were approximately 3 months of age. Effects of birth weight and gender were analysed using a mixed model ANOVA. LBW piglets showed an impaired reference memory performance ($F=5.16$; $df=1,360$; $P=0.024$) compared to NBW piglets throughout the acquisition phase. No effect of birth weight was found on working memory scores. Birth weight did not influence the latency to visit the first hole of a trial, indicating comparable motivation to perform the task for LBW and NBW piglets. However, LBW piglets were on average slower to find their first reward ($F=8.54$; $df=1,360$; $P=0.004$) than NBW piglets. Gender did not affect any of these measures. Together, these results confirm our hypothesis that LBW causes cognitive deficits in pigs. Furthermore, male and female piglets show comparable cognitive performance in the acquisition of the holeboard task. This suggests the negative cognitive effects of LBW are not gender-specific.

Never wrestle with a pig

Dorte Bratbo Sørensen
University of Copenhagen, Department of Veterinary and Animal Sciences, Grønnegardsvej 15,
1870 Frederiksberg C, Denmark; brat@sund.ku.dk

The use of positive reinforcement training (PRT, 'clicker training') has slowly spread from the world of zoos and pets to laboratory animal facilities housing non-human primates, chimpanzees or dogs. Especially in apes and non-human primates, the power of PRT to decrease stress and facilitate veterinary procedures has been demonstrated, but similar studies are lacking in pigs. Compared to primates, the pig is a much more commonly used lab animal, but even though the pig is easy to train, often only simple habitation- and luring techniques (if any) are used to reduce stress and fearfulness in the pigs. According to the EU directive (2010/63/EU Directive on the protection of animals used for scientific purposes; preamble 31), animal welfare should be given the highest priority. Hence, training of laboratory animals should always be implemented to the highest possible extent. This presentation will provide an overview of basic positive reinforcement training principles and pit-falls and examples of individual effects of PRT on laboratory pigs. We have used simple PRT such as target training to rapidly and effectively gain benefits such as improved human-animal contact, less fear of the animal caretaker and a quicker and smoother performance of e.g. weighing of the pigs. Moreover, by implementing PRT in daily management, we simply add control, predictability and positive anticipation to the environment of the pigs; features which are all said to increase animal welfare. This presentation will discuss the choice of training method (PRT, luring, habituation, etc.) in relation to the purpose of training and the importance of competences and engagement of the trainers and the equipment used. Also the time allocated for training and initial socialization to humans needs to be considered, as pigs traditionally are used for short-term studies, where time for training is seldom included. This lack of time for optimization of training, however, seems not to be in accordance with the EU directive and hence there is definitely room for improvement. Videos demonstrating training techniques and training outcomes when using simple luring compared to PRT are presented and discussed as are welfare gains and economic/other benefits of training laboratory pigs.

Exploring the effect of automated milk feeding stall design on dairy calf behavior

Tanya Wilson, Stephen Leblanc, Trevor Devries and Derek Haley
University of Guelph, Population Medicine, 50 Stone Rd E, N1G 2W1, Canada;
tdevries@uoguelph.ca

There is evidence that the health and welfare of young dairy calves may be improved by increasing milk allowance and by providing milk through a teat. These aspects are easily incorporated into automatic milk feeding systems, which promote group housing that has been demonstrated important for social development of calves. Little is known about how calves interact with automatic milk feeders. We investigated the effect of stall design features on calves learning to use the automatic feeder. Sixty-six male and 53 female Holstein calves were enrolled at 4 d of age and introduced to a group pen and trained on an automatic feeder; they were allowed to suck on a trainers' fingers and guided to the teat. Calves were allocated to 1 of 2 different stall designs: one with gated (open) side walls (n=59), the other with solid side walls (n=60). Over a 72-h period after training on the automatic feeder, data from the automatic feeder was collected and calf behaviour was monitored by video. The hypothesis for this study was that solid stalls would result in a longer latency to approach and feed from the automatic feeder. Main outcomes measured were latency to first voluntary visit to the feeder, latency to first feeding, time spent in feeder, amount of milk drank, and exploratory behaviour such as sniffing and licking of the feeder. Data were analyzed using a mixed-effect linear regression model. The overall latency for calves to first voluntarily drink from the feeder was 5.3 h greater (P=0.05; SE=2.7) with a gated side compared to solid. The average time taken to first drink was 23.7 h (min 2.9, max 69.1) with gated sides and 15.1 h (min 3.2, max 54.5) with solid sides. Male calves also drank 2.0 l more (P=0.05; SE=1.0) milk over the 72-h period than females. Results from this experiment show that simple features of a stall on an automatic milk feeder can influence dairy calf behaviour. A shorter latency to first voluntary milk feeding from the automatic feeder with solid sides compared to gated, suggests simple differences between automatic feeders can influence calves feeding behaviour, and should therefore be taken into account by producers and companies manufacturing the equipment.

Poor reversal learning in dairy cows is associated with endometritis risk after calving

Kathryn Proudfoot[1,2], Becca Franks[2], Alexander Thompson[2] and Marina Von Keyserlingk[2]
[1]The Ohio State University, Veterinary Preventive Medicine, 1920 Coffey Rd, Columbus, OH 43210, USA, [2]The University of British Columbia, Animal Welfare Program, 2357 Main Mall, Vancouver, BC, V6T 1Z4, Canada; proudfoot.18@osu.edu

Intensively housed dairy cows face a number of stressors at calving, and are at high risk of disease. How individual cows cope with these stressors may influence their health, but little research has identified characteristics of at-risk cows. Individual differences in neuroticism have been found to be associated with disease risk in other species; neuroticism is characterized by a prolonged negative affective state, vulnerability to stressors, and cognitive deficits including poor 'reversal learning' or the inability to change from a previously-rewarded behaviour to a new behaviour. Our aims were to determine if dairy cows that perform poorly on reversal learning tasks before calving have a higher risk of endometritis after calving. Cows were cared for according to a protocol approved by the UBC's Animal Care Committee. Four weeks before calving, cows were assigned to one of 7 groups (4 cows/groups, n=28) and moved into a free stall pen with 4 electronic feed bins, 8 lying stalls, and 4 non-experimental cows each assigned to a single feed bin throughout the experiment. Feed bins recorded feed intake during each visit and the number of times a cow tried to access a bin that she was not assigned to. On the first day, cows were assigned to have access to one feed bin that they had to share with a non-experimental partner. For the next 4 weeks, cows were assigned to a new bin and a new partner every other day, exposing them to continuous reversal-learning challenges until calving. To determine if cows showed stable individual differences in reversal learning, we collected two variables on days that cows were assigned to a new bin: (1) the number of times a cow tried to visit the bin she had been assigned to the previous day after successfully feeding from her new bin, and (2) the number times a cow visited any bin after successfully feeding from her new bin. Visits to the previous rewarded bin were divided by total visits (rewarded and unrewarded) to determine the proportion of time a cow persisted at her previous bin. Endometritis (>5% neutrophils) was diagnosed using a uterine cytology smear 3-5 wk post-calving. Cows showed stable individual differences in their persistence in attempting to visit their previous, now unrewarded, bins: the correlation between persistence during the first and last two weeks was 0.55 (P<0.01). There was no correlation between poor reversal learning and feed intake (P>0.6). However, logistic regression revealed that the most persistent cows had an increased probability of developing endometritis (P<0.04). In conclusion, some cows were consistently poor at changing a previously-rewarded behavior, and these cows were more likely to develop endometritis regardless of feed intake. The results provide evidence that one characteristic of neuroticism is associated with endometritis risk in dairy cows; more research is needed to determine the relationship between other characteristics of neuroticism and health in dairy cows.

Artificial intelligence algorithms as modern tool for behavioural analysis in dairy cows

Håkan Ardö[1], Oleksiy Guzhva[2], Mikael Nilsson[1], Anders Herlin[2], Lena Lidfors[3] and Christer Bergsten[2]
[1]Lund University, Centre for Mathematical Sciences, Sölvegatan 18A, Box 118, 221 00 LUND, Sweden, [2]Swedish University of Agricultural Sciences, Biosystems and Technology, Sundsvägen 16, Box 103, 230 53 Alnarp, Sweden, [3]Swedish University of Agricultural Sciences, Animal Environment and Health; Applied Ethology Unit, Gråbrödragatan 19, Box 234 532 23 Skara, Sweden; oleksiy.guzhva@slu.se

Behaviour is an invaluable tool for monitoring health and welfare of dairy cows. While the behaviour of an individual has a direct correlation with actual health and stress levels, it is also influenced/influences behaviour of other individuals within the group. In dairy cows, group hierarchy and group dynamics are essential indicators of cow's performance. Correct classification of behaviours occurring within the group is of importance as some interactions are positive, and strengthen the bond between cows and others are possibly deleterious and can reflect a struggle for resources. Therefore, gaining knowledge about dyadic relationships within the group could help in creating an animal-friendly environment and reduce stress related to competition. The aim of this study was to take the first step towards an automated system for behavioural analysis. The complete dataset contained 4 months, from two different seasons, of recordings (400 million frames, among which 2,148 were randomly selected and annotated) collected from 3 cameras installed in a holding pen of a conventional dairy barn with 252 Swedish Holstein cows. Then an algorithm removed uninteresting parts (e.g. those, containing behaviours that were not relevant or without animals in the scene) of the recorded video material. The annotated subset contained in total 9,278 cows. A simple 'watchdog' algorithm extracting frames containing two or more cows was implemented, and it was possible to discard 38% of the recordings as uninteresting. To remove more of the uninteresting video, additional features such as the minimum distance (as cows need to be close to interact) between the cows in the scene and duration of contact, were also extracted. Then a short sequence containing interactions consisting of 187 frames evenly sampled over 10 minutes was extracted and manually annotated by an expert to be then used for cross-validation with automatically annotated data. Five different interactions (used in previously developed 7-point shape model) were manually identified as relevant for this study: body pushing, butting, head butting, head pressing and body-sniffing. Frames where any of these interactions were present, were considered interesting and all other frames uninteresting. Most, 97.1%, of the images were perfectly interpreted, i.e. all cows present were detected with no extra detections, with false alarm rate of 2.9%. The proposed algorithm was able to discard 50% of uninteresting video material, while only losing 4% of potentially interesting fragments. This, in combination with features for proximity analysis, allows to conclude that modern computer algorithms could help in assessing group structure in dairy cows as well as save valuable expert time.

Best friends forever, new approach to automatic registration of changes in social networking in dairy cows

Per Peetz Nielsen[1], Luis E.C. Rocha[2], Olle Terenius[3], Bruno Meunier[4] and Isabelle Veissier[4]
[1]University of Copenhagen, Department of Veterinary and Animal Sciences, Grønnegårdsvej 8, 1879 Frederiksberg C, Denmark, [2]Karolinska Institutet, Department of Public Health Sciences, Solnavägen 1, Solna, 171 77 Stockholm, Sweden, [3]Swedish University of Agricultural Sciences, Department of Ecology, P.O. Box 7044, 750 07, Sweden, [4]UCA, INRA, VetAgro Sup, UMR1213 Herbivores, 63122 Saint-Genes-Champanelle, France; ppn@sund.ku.dk

Until recently, it was not possible to record social network in dairy cows over a long period. However, thanks to the development of systems using ultra-wideband for real time location, it is now possible to follow the exact position of a large group of cows over an unlimited period. The aim of this study was to examine the changes in social contact that occurs when cows are removed from a stable group and at the same time adding new cows into the remaining group. We used a position system using UWB to record x-y position of dairy cows at the INRA research farm (UE1414 Herbipole) in Marcenat, France. The tag was placed on the neck of the cow. This technology enables us to look more deeply into the social network and structure of dairy cows and examine how disturbances affect this structure. Furthermore, it can be used to develop automated systems for the detection of dairy cows at risk of being infected by other cows and for the reduction of the spread of infection in the herd, with the information on which cows that have been exposed to contaminated areas. A social network was defined by a group of nodes, representing the cows, connected by links, representing the social interactions between the cows. Every second two cows were within a radius of 1.25 m of each other, we considered it as a link. Such links were defined only for cows outside the resting/cubicle area. The value of 1.25 m corresponds approximately to twice the distance between the neck and the nose of the cow and thus captures potential face-to-face interactions. We aggregated all links during the day and assigned the total time to the aggregated link between the respective cows. The connectivity (or node degree) of a cow to another cow was defined as the total time exceeding 1,500 seconds of connections during a day. The data consisted of a group of 26 cows where 9 cows were moved and after that, 8 different cows were added into the group. Sociograms (a graphic representation of social links that an individual has) of the links between the cows showed that the first seven days after the new cows are added into the group, there were many social links, but these links changed each day. The number of average connections during the week that exceeded 1,500 seconds per day was smaller in the new constellation and mainly occurred between the original cows in the group. We interpret this as an increased connectivity due to a build-up of a new hierarchy. It is furthermore concluded that cow-tracking systems have a huge potential for examination of the social structure and changes of these structures in dairy cows. Further studies are needed with more groups of cows in order to conclude how the hierarchy develops over time in new constellations of dairy cows.

Phenolab: ultra-wide band tracking shows feather pecking hens spent less time in close proximity compared to controls

Elske N. De Haas[1], Jerine A.J. Van Der Eijk[1], Bram Van Mil[2] and T. Bas Rodenburg[1]
[1]Wageningen Universit & Research, Behavioural Ecology Group, Adaptation Physiology Group, De Elst 1, 6708 WD Wageningen, the Netherlands, [2]Noldus Information Technology B.V., Nieuwe Kanaal 5, 6709 PA Wageningen, the Netherlands; elske.dehaas@wur.nl

Hens which feather peck can cause multiple victims in a group. As a result pen-mates could keep a greater distance from feather peckers. Our aim was to relate time spent in close proximity between birds selected divergently on feather pecking. Automatic ultra-wide band location data was obtained by using an active sending tag (Ubisense®) placed in a backpack on the birds. A group of birds was then placed for 15 minutes in a barren test-room equipped with four receiving beacons. We used 37-week old White Leghorn laying hen lines selected for high (HFP) or low feather pecking (LFP), and an unselected control (CON) line (n=76 in total, 5-7 hens per pen; 4 pens per line). Hens were habituated to the backpacks four weeks prior to observations, and were tested for 5-min to the test-room individually one week prior to group testing. Sample rate of tags was set to twice per second, detecting location of each tag separately and simultaneously. Location data was provided by TrackLab software (Noldus, Wageningen, The Netherlands) and compared between tags by Excel calculations. Percentage of time in close proximity was defined as being less than 25 cm apart from another tag, calculated for each bird and averaged per group. Behavioural sampling of feather pecking of 2×20 min at 28-29 wks of age was included on pen-level and used for identification of peckers (>2 bouts/20 min). Percentage of time in close proximity was correlated to feather pecking (pearson correlation), tested to differ between peckers and average of the group (t-test of groups with peckers) and tested to differ between lines (ANOVA including the number of animals per pen as covariate). Feather pecking on pen-level was not correlated to proximity measures (r=-0.28, P>0.10), but birds spent less time in close proximity of the feather pecker as opposed to time spent on average with other pen-mates (17.6±1% vs 21.6±1.2, t16=-1.18, P=0.04). Hens of the high feather pecking line spent a lower percentage of time in close proximity compared to controls (HFP: 18.9±2.2% vs CON: 39.8±10.1% vs LFP: 27.0±1.6%; $F_{2,12}$=5.4, P=0.05). The number of hens per group did not influence proximity ($F_{1,12}$=0.39, P=0.93). This data shows that selection on feather pecking has affected proximity. Less time in close proximity in HFP vs LFP groups and with feather peckers vs non-peckers could be explained by higher activity of peckers and in HFP groups. As feather damage was pre-existing in HFP birds, proximity differences between HFP and LFP groups could not be used as predictors for feather pecking. Whether this pattern of proximity is linked to the divergent genetic selection or a response to the actual occurrence of feather pecking will be further examined on group level in other home-pen situations. This automatic tracking method seems promising to pick up within group differences in behavioural patterns and might ultimately be used to predict feather pecking in laying hens.

Floor feeding over multiple drops does not increase feeding opportunities for submissive sows.

Natalia Zegarra[1], Paul. H. Hemsworth[1] and Megan Verdon[2]
[1]*The Animal Welfare Science Centre, The University of Melbourne, Parkville, Victoria, 3010, Australia,* [2]*Tasmainian Institute of Agriculture, The University of Tasmainia, Burnie, Tasmania, 7320, Australia; megan.verdon@utas.edu.au*

Feeding sows from the floor is one of the simplest and cheapest method of feed delivery, but often results in high levels of aggression as sows compete for access to food. This may lead to subordinate sows sacrificing the opportunity to feed in an attempt to avoid receiving aggression. This study examined whether floor feeding sows their daily ration over multiple bouts per day creates more opportunities for subordinate sows to feed. Two-hundred pregnant gilts were mixed into pens of 10 (1.8 m²/gilt) within 7 days of insemination and floor fed four times per day (07:30, 09:00, 11:00, 15:00 h). The identity of gilts that were feeding (head down, rooting the ground/feed, chewing) was recorded, along with her location in the pen [on the cement area where feed was scattered (i.e. feeding area; F), directly under the feeder (i.e. high feed availability; HF), on the slats at the back of the pen (i.e. little or no feed availability; NF)], every 30 s for 20 min after each feed drop on days 2, 9 and 51 post-mixing. Gilts were also classified as dominant (D), subdominant (SD) or submissive (S), as described elsewhere. GLMM in conjunction with LSD tests were used to analyse the effects of gilt classification, day and feeding bout on the time gilts spent feeding in each location. Each model included a repeated effect of day and feeding bout within gilt, and pen nested within replicate as a random effect. Regardless of feeding drop (P>0.05) or day (P>0.05), D spent the most, and S the least, time feeding at the HF location. In feed drops 1, 2 and 4, the time SD spent HF was intermediate to S and D, but in drop 3 the time SD spent HF was comparable to S (classification × drop interaction P=0.01). Irrespective of feed drop (P>0.05), S gilts spent less time in the feeding area (P<0.001), and more time in the NF location (P=0.03), than D and SD, while the latter two did not differ. Feeding gilts over multiple drops does not increase feeding opportunities for subordinate sows. Regardless of feed drop or day, D gilts occupied the area of high feed availability and likely had the highest feed intake. That SD appear to adopt a strategy of feeding between and around the D gilts in areas of reduced feed availability. However, S gilts appear to be at greatest risk of low feed intake, which may result in reduced body condition and increased hunger and frustration. Further research is required to determine how the feed intake of subordinate sows can be increased in floor feeding systems, while also protecting these animals from aggression.

The effect of socialization before weaning on social behaviour and performance in piglets (Sus scrofa) at a commercial farm

Laura Salazar[1], Irene Camerlink[2], Heng-Lun Ko[3], Chung-Hsuan Yang[1] and Pol Llonch[3]
[1]Edinburgh University, Royal (Dick) School of Veterinary Studies, Easter Bush, Midlothian, EH25 9RG, Edinburgh, United Kingdom, [2]Scotland's Rural College, Animal and Veterinary Sciences Research Group, West Mains Rd., EH9 3JG, Edinburgh, United Kingdom, [3]Universitat Autònoma de Barcelona, School of Veterinary Science, Bellaterra, 08193, Cerdanyola del Vallès, Spain; laurasalazarhofmann@gmail.com

Although positive effects of socializing piglets pre-weaning on behaviour and performance have been confirmed, it has not been proven in commercial conditions. The aim of the present study was to investigate behaviour (locomotor play, aggressive and social behaviour), aggressions (skin lesion score) and performance (weight and average daily gain (ADG)) of piglets socialized at different ages during lactation at a large scale commercial farm. We hypothesized that socialization increases social behaviour during lactation, reduces skin lesion score (less aggression) and results in improved weight gain after weaning. After approval from IACUC, piglets were either not socialized (CON; n=12 litters) or socialized at 7 (M7; n=20 litters) or 14 days of age (M14; n=20 litters) by removing the separating barrier between two adjacent pens. At weaning (day 25), piglets were regrouped (~39 pigs/group, 8 groups in total) with unfamiliar piglets of the same treatment. Social, aggressive and locomotor play behaviour was measured using instantaneous scan sampling on day 7, 14, and 21 in all treatment groups. Skin lesions were counted before (-1) and after (+1 and +2) mixing at socialization and weaning. Weight was recorded weekly from one day after farrowing until ten days after weaning (day 35). Data were analysed using mixed models accounting for sow and pen effects. Socialized (M7 and M14) piglets showed three times more play behaviour (1% of observations) compared to CON piglets (0.2% of observations) on day 14 (P=0.03). At the time of socialisation, aggression was more often observed in M14 (1.5%) than in CON (0.6%; P=0.04) and tended to be higher in M7 (1.4%) than in CON (P=0.07). Skin lesion score increased after socialisation in M7 (before: 0.40±0.09; after: 1.53±0.20; P<0.001), but not in M14 (before: 0.82±0.12; after: 0.80±0.13; P=0.68). After weaning, skin lesions increased in CON (before: 0.21±0.06; after: 1.03±0.21; P<0.001), but not in M7 (before: 1.11±0.39; after: 0.74±0.21; P=0.007) neither M14 (before: 0.97±0.33; after: 1.20±0.32; P=0.44). Treatment groups did not significantly differ in birth weight (CON 1.7±0.03; M7 1.5±0.03; M14 1.5±0.03 kg; P=0.41) but from day 7 to 21 the control group was heavier than socialized piglets (P<0.05). Although the decline in ADG after weaning was not significantly different between treatments (CON 0.093±0.016; M7 0.082±0.012; M14 0.093±0.011 kg/d) (P=0.68), the final weight (day 35) was not different among treatments (CON 7.04±0.54; M7 6.07±0.44; M14 6.40±0.44 kg) (P=0.27). In conclusion, in a commercial farm, socialization can increase social behaviour during lactation and reduce aggression when mixing with unfamiliar piglets at weaning without reducing performance, which is in line with research outcomes under controlled experimental facilities.

Socialisation, play behaviour, and the development of aggression in domestic pigs (*Sus scrofa*)

Jennifer Weller[1], Simon Turner[2], Irene Camerlink[2], Marianne Farish[2] and Gareth Arnott[1]
[1]Queen's University Belfast, School of Biological Sciences, Queen's University Belfast, Belfast, BT7 1NN, United Kingdom, [2]Scotland's Rural College (SRUC), Animal Behaviour and Welfare, RSUC Roslin Institute, EH25 9RG, United Kingdom; jweller01@qub.ac.uk

Aggressive behaviour post-weaning in domestic pigs is a major welfare concern that remains to be fully addressed. The role of play behaviour in the development of aggression has received little attention, yet it has been suggested that early life social play facilitates the development of later life social skills, including those used to resolve agonistic encounters. We hypothesise that socialising piglets during early life will increase the opportunity for play experience, and that individuals that have experienced more play will be less aggressive with conspecifics during later life. Additionally, we examine sex differences in play behaviour, making connections to later life aggression and expected social roles. This study was conducted using 196 piglets from 19 litters. Of these, 8 litters were selected to undergo socialisation and were allowed to interact freely with the individuals of a neighbouring pen between days 14-26 after birth. The remaining 11 litters acted as controls and remained isolated within their litter groups until post-wean mixing. In both socialised and control pigs, all occurrences of social play with other piglets (play fighting, nudging and chasing) and sows (naso-naso contact and climbing) were recorded over a 6 hour period for 5 days that spanned the socialisation period. At 7 weeks old, piglets (n=109) from both treatments had their aggressiveness assayed using the previously validated resident-intruder (RI) test, repeated with different intruders on two consecutive days. Latency to attack was recorded, or the interaction terminated if no attack occurred within five minutes. All work conformed to required legislation, receiving ethical approval and adhered to ISAE guidelines. Linear mixed effects models revealed that socialisation did not significantly increase total play fighting behaviour (P=0.83) or play behaviour as a whole (P=0.33). Despite this, socialised piglets were found to have a shorter attack latency than controls (P=0.04, Soc=258.9 s, SE=±23.5 s, Con=384.3 s, SE±27.0 s) suggesting that some other aspect of socialisation influences aggression. Controls were found to interact with the sow significantly more than socialised individuals (P=0.02, Con=13.4, SE±0.9, Soc=6.8, SE±0.5) and there was a positive correlation between sow-aimed play and attack latency (P=0.04, rho=0.19). Regarding sex differences, males engaged in more play fighting than females (P<0.001, Male=27.0, SE±1.7, Female=18.3, SE±1.4) although females showed a shorter attack latency than males (P=0.0003, Female=261.4 s, SE±28.9 s, Male=352.2 s, SE±23.6 s). Thus, early life socialisation resulted in increased aggressiveness in the context of an RI test, possibly reflecting an increased ability to rapidly establish a dominance hierarchy. This assertion will be explored, while also examining sex effects, in due course by considering the effects of socialisation on the assessment strategies used during contests and mixing practices in later life.

Intensity of aggression in pigs depends on their age and experience at testing

Irene Camerlink[1], Marianne Farish[1], Gareth Arnott[2] and Simon P. Turner[1]
[1]Scotland's Rural College (SRUC), Animal Behaviour & Welfare, Animal Veterinary Sciences Research Group, West Mains Rd., EH9 3JG Edinburgh, United Kingdom, [2]Queen's University, School of Biological Sciences, Institute for Global Food Security, University Road, BT7 1NN Belfast, United Kingdom; irene.camerlink@sruc.ac.uk

Type and frequency of behaviour in any species is strongly dependent upon the situation in which it is measured. Our aim is to illustrate, using the example of aggression, how research outcomes can change due to age and experiences of the animals at testing. Data from three experiments on pig aggression (approved by ethical committees) were combined. All trials included dyadic contests with the same test set-up, carried out on the same farm. Within and between tests, pigs varied in age and experience before the contest. Pigs were either tested at 8, 10 or 13 weeks of age and had either never encountered an unfamiliar pig (E0, $n=550$ pigs of 8 and 10 wk of age), had been co-mingled with another litter pre-weaning for 2 wks (EL, $n=184$; 8 wk age), had encountered a single unfamiliar pig at 10 wk age (E1, $n=136$; 13 wk age) or had been subject to group mixing at 11 wks of age (EM, $n=168$; 13 wk age). Comparisons for age were made between E0-8 and E0-10 whereas the effect of experience was compared between E0-8 vs EL-8 and E1-13 vs EM-13. For the contest, two unfamiliar pigs simultaneously entered an arena (2.9×3.8 m). Contest duration, skin lesions as a result of bites, and agonistic behaviour were recorded. Contests lasted max 30 min and had strict end-points to prevent lasting harm. Data were analysed using mixed models that accounted for trial and pen in the random effects. Contests with an escalated fight were longer (fight 287 s; no fight 158 s; $P<0.001$) and fight occurrence was thus included as a fixed factor. Contest duration increased with age, with naïve pigs (E0) on average spending 164 s in contest at 8 wk of age and 240 s at 10 wk of age ($P<0.001$). Experience of co-mingling decreased the contest duration at 8 wk of age from 253 s (E0) to 202 s (EL; $P<0.01$). In contrast, experience of group mixing at 10 wk of age increased the contest duration at 13 wk of age from 199 s (E1) to 299 s (EM; $P<0.001$). Skin lesions (overall average of 45 lesions) increased with contest duration ($P<0.001$). When corrected for contest duration, the number of lesions were unaffected by age when comparing pigs of 8 vs 10 wk age ($P=0.52$), and were unaffected by experience (E0-8 *vs* EL-8: $P=0.11$; E1-13 *vs* EM-13: $P=0.50$). Ritualized display behaviour, such as foaming and piloerection, increased with age, but almost exclusively in males. Foaming was performed by 19% of the males when 8 wk of age, by 26% when 10 wk of age, and by 68% when 13 wk of age. In females the percentage was 5, 5, and 15%, respectively. This highlights strong age related sex differences even at a pre-pubertal age. We recommend that the age and experience of test animals are carefully considered prior to conducting behavioural research, in particular for research on aggressive behaviour.

Effect of overstocking during the dry period on the behaviour and stress level of dairy cows, and on pre-weaning calf growth

Mayumi Fujiwara[1,2], Kenny Rutherford[2], Marie Haskell[2] and Alastair Macrae[1]
[1]*University of Edinburgh, Royal (Dick) School of Veterinary Studies, Easter Bush, Midlothian, EH25 9RG, United Kingdom, [2]Scotland Rural College, West Mains Road, Edinburgh, EH9 3JG, United Kingdom; mayumi.fujiwara@sruc.ac.uk*

On commercial dairy farms, dry cows are usually kept in dynamic social groups. Limited access to resources such as feed and/or lying space is known to increase competition and alter feeding behaviour in dairy cows. Stressful maternal experience can have a negative impact on fetal development, which could affect the offspring throughout its postnatal life. The aim of this study was to investigate the effects of overstocking during the dry period on the behaviour and welfare of dairy cows, and to investigate the effects of maternal social experience on calf bodyweight and growth. Forty-eight Holstein cows (calving over a six month period) were dried off 8-9 weeks before their expected calving date, and kept in cubicle sheds with a yoke feed barrier during the first six weeks (far-off) and then kept in straw yards with a post-and-rail feed barrier until calving (close-up). Cows were allocated to either high (H) or low (L) stocking density groups after dry-off. H and L cows had 0.5 vs 1.0 yokes/cow and 1.0 vs 1.5 cubicles/cow respectively in the far-off period, and 0.3 m vs 0.6 m feed face/cow and 6 m^2 vs 12 m^2 lying space/cow respectively in the close-up period. The time spent feeding during the 3 hours after feed delivery was recorded three days per week, and agonistic social interactions were observed two days per week (0-20 min, 40-60 min and 80-100 min after feed delivery). Faecal glucocorticoid metabolites (FGCM) were measured during the dry period at 5 time-points. Bodyweight of calves (n=44) was measured at birth and at weaning (at 49 days old), and daily gain from birth to weaning was calculated. Data were analysed by Residual Maximum Likelihood (cow data) or General Linear Model (calf data) using GenStat following log transformation where necessary. Means and standard errors are presented, and means obtained from the transformed data were back-transformed to the original scale, and 95% confidence intervals (CIs) are reported. H cows spent less time feeding (72.1±4.4 min) than L cows (86.8±4.8 min) during the 3 hours after feed delivery (W=5.5, P=0.023), but engaged in more agonistic interactions (H: 3.45/group, 2.92-4.05; L: 1.93/group, 1.57-2.34). The concentration of FGCM was not affected by treatment, but FGCM during the dry period were significantly higher than that at dry-off (dry-off: 184.1 ng/g, 80.1-423.1; week2: 379.3 ng/g, 161.1-893.1; week5: 323.6 ng/g, 140.8-743.7; week6: 368.1 ng/g, 160.1-846.1; pre-calving: 357.3 ng/g, 150.0-851.2, W=85.6, P<0.001). No effect of maternal treatment was found in the calf bodyweight at birth and weaning, or in the pre-weaning daily gain. Overstocking of dry cows resulted in a shorter feeding time during the 3 hours post feeding and more frequent competition at the feed face. However overstocking did not affect the concentrations of faecal glucocorticoid metabolites. Maternal competitive experience and altered feeding behaviour during the dry period did not have negative consequences on calf bodyweight or pre-weaning growth.

Relationship between social rank and reproductive, body and antler traits of pampas deer (*Ozotoceros bezoarticus*) males

Matías Villagrán[1], Florencia Beracochea[1], Luděk Bartoš[2] and Rodolfo Ungerfeld[1]
[1]Facultad de Veterinaria, Universidad de la República, Departamento de Fisiología, Lasplases 1620, 11600, Uruguay, [2]Institute of Animal Science, Prague, Department of Ethology, Přátelství 815, 104 00, Czech Republic; matiasv06@gmail.com

The aim of this work was to determine the relationship between social rank and parameters of body size, antler size, testosterone concentration, and semen characteristics of pampas deer males. The study was performed in a semi-captive population of pampas deer allocated in the Estación de Cría de Fauna Autóctona Cerro Pan de Azúcar (Uruguay), in February and March (early to mid-rut). Data was collected from five groups of 4-7 adult males (2-7 y old) in four different seasons. Overall, 17 animals were used, but as 6 of them were allocated in different groups in different years, the total number of data considered for the analysis was 26. The agonistic behavior was recorded during the rut identifying the dominant and the subordinate animal in each interaction. The individual Success Index (SI) was calculated as the number of individuals displaced/(number of individuals displaced + number of individuals that displaced it) (range = 0 to 1). Animals were anesthetized, and blood was collected to measure testosterone concentration. Body weight, neck perimeter, and shoulder height, circumference of the antlers' coronet, length of second and third tines and total antler length were measured. Semen was collected by electroejaculation of the anesthetized animals. Semen volume, semen concentration, and percentages of spermatozoa with progressive motility, normal morphology, membrane integrity, and acrosomal integrity were determined. The total number of spermatozoa of each variable was calculated (percentage of the variable × total number of spermatozoa). Applying a principal component analysis, following components were created: semen vitality (variables: total number of spermatozoa with progressive motility, membrane integrity, and normal morphology), acrosome integrity (percentage of spermatozoa with intact acrosome), antlers (circumference of the coronet, and length of the second and third tine), and body size (neck perimeter, body weight, shoulder height). The relationship between dependent variable SI and age and the components was determined using a generalized linear mixed model designed for repeated measurements. Success Index was basically dependent on age [$F(5, 7.7)=10.0$; P=0.003] and body size [$F(2, 12.7)=5.7$; P=0.02]. In 1-4 y old males, males with higher SI had greater *body size*, while in fully mature, 5-7 y old males increasing body size was observed with decreasing SI. Social rank tended to be associated with greater values of antlers component [$F(1, 10.3)=3.9$; P=0.07] but not on testosterone concentration (P=0.16). The semen vitality component was positively associated with SI [$F(1, 2.7)=44.9$; P=0.01] while the acrosome integrity was not related to SI at all (P=0.34). In conclusion, while social rank of pampas deer males allocated in small groups in semi-captivity was positively associated to semen vitality and probably to antlers size, in younger individuals SI was positively associated to body size.

Rearing bucks (*Capra hircus*) isolated from females affects negatively their sexual behavior when adults

Lorena Lacuesta[1], Julia Giriboni[1], Agustín Orihuela[2] and Rodolfo Ungerfeld[1]
[1]Facultad de Veterinaria/Universidad de la República, Departamento de Fisiología, Lasplaces 1620, 11600, Uruguay, [2]Facultad de Ciencias Agropecuarias/Universidad Autónoma de Morelos, Cuernavaca, 62209, Mexico; lacuesta16@gmail.com

In some domestic ruminants, contact with females is necessary for normal development and display of sexual behavior. The aims of this experiment were to determine if rearing bucks isolated from females affects negatively their sexual behavior when adults, and if this negative effect is overcome after four short-term contacts with females. Sixteen Saanen male kids were maintained during one year in two groups: kids reared in permanent direct contact with four adult goats (FEM; n=7), and kids that remained isolated from females (ISO; n=9). When bucks were 12 mo-old, females were removed and both groups were joined in the FEM pen. Nine months later all bucks were individually exposed four times to unrestrained estrous females in 20 min tests, each performed 10 days apart, recording courtship (sum of anogenital sniffs, flehmens and lateral approaches) and mounting behaviors (sum of mount attempts, mounts without and with ejaculation) and the latency to the first ejaculation. Data were compared with the glimmix proc assuming a Poisson distribution of data and the combination of the number of bucks that ejaculated and the latency at which they did it was compared with a survival analysis. The number of courtship behaviors was greater in FEM than ISO bucks ($P<0.0001$), varied with number of test ($P<0.0001$) and there was an interaction between treatment and number of test ($P<0.0001$) (test 1: 75.7±16.3 vs 45.5±17.9; 2: 86.4±24.6 vs 43.4±25.7; 3: 44.3±6.7 vs 48.0±21.6; 4: 51.1±15.2 vs 29.1±8.5, FEM and ISO respectively). The interaction was explained by a greater number of courtship events in FEM than in ISO bucks in the first, second and fourth tests ($P<0.0001$ for all these tests). Bucks reared with females displayed more mounts with ejaculation and total number of mounts than ISO bucks (mounts with ejaculation: 3.2±0.9 vs 1.5±0.5, FEM and ISO respectively; total mounts: 3.6±0.9 vs 1.75±0.6, FEM and ISO respectively, $P<0.0001$ for both). Mounts with ejaculation and total number of mounts did not vary according to the number of test and there was not interaction between treatment and number of test. The combined effect of number of bucks that ejaculated and the time at which they first ejaculated (survival analysis) favored FEM bucks in the first and second tests (test 1: 7/7 and 46.0±15.3 s vs 5/9 and 159.0±109.6 s; test 2: 7/7 and 43.8±11.5 s vs 5/9 and 196.0±159.0 s, FEM and ISO respectively; $P<0.03$). We concluded that the lack of contact with females during the rearing period affects negatively adult bucks' sexual performance, effect that cannot be overcomed with at least four repeated exposures to estrous does. In bucks, contact with females during the rearing period leads to develop normal male sexual behavior. As goats kids are reared in females groups, this allows bucks to mate females effectively at an earlier age and thus, increase the reproductive success of the specie.

Social behavior and adrenocortical activity in goats up to four wees after grouping

Susanne Waiblinger[1], Eva Nordmann[1] and Nina Maria Keil[2]
[1]*Institute of Animal Husbandry and Welfare, University of Veterinary Medicine Vienna, Veterinärplatz 1, 1210 Wien, Austria,* [2]*Center for Proper Housing of Ruminants and Pigs, Agroscope Reckenholz-Tänikon, Federal Veterinary Office, Agroscope Reckenholz-Tänikon ART, Research Station ART, Tänikon, 8356 Ettenhausen, Switzerland; susanne.waiblinger@vetmeduni.ac.at*

Changing group composition is frequently practiced in farm animals and known to be stressful. Most studies in goats investigated integration of few animals in a larger group for only a short time. This study investigated the course of agonistic social interactions, adrenocortical activity and injuries up to week 4 after merging two groups. Seven groups of 7 and one group of 6 female dairy goats (n=55, 3-8 years old) were grouped together to form 4 groups of 14 or 13 animals. Two groups consisted only of horned goats, the other 2 only of hornless (polled or dehorned) for experimental reasons. The groups were housed in pens with 2.2 m^2/goat on deep litter with an elevated feeding place. For adrenocortical activity, faecal cortisol metabolites (FCM) were measured in samples taken rectally on 2 days per period: 2 weeks before grouping (Before), after grouping on day1 and 2 (Week1), 2 weeks after grouping (Week2), 4 weeks after grouping (Week4). Observations of agonistic interactions by continuous behaviour sampling were conducted in Week1 to 4. Initiator and receiver of agonistic interactions were identified individually. Number of agonistic interactions in total (AgoTotal), with body contact (AgoBodyC) and without body contact (AgoNoBodyC) were calculated per animal per 2.25 h. Goats were examined for injuries once per period. Behaviour (log-transformed) and FCM were analysed by mixed models with animal nested in group as random factor; injuries are depicted descriptively. Adrenocortical activity increased from Before (261 ng/gr) to Week1 (312 ng/gr) and then decreased (Week2: 245, Week4: 240 ng/gr; main effect period: P<0.001, $F_{3,123}$=6.59). In contrast AgoTotal and AgoBodyC peaked in Week2 (model estimate: hornless: 0.833, horned: 0.620). In Week4 the number of these interactions was still higher (model estimate: AgoTotal: 0.744; AgoBodyC: 0.517) than in Week1 (0.547; 0.440) for hornless goats, but lower for horned goats that also in general were on a lower level than hornless goats (Week4: 0.481; 0.240; Week1: 0.520; 0.381; interaction period×horn status: P=0.042, $F_{2,105}$=3.26 and P=0.054, $F_{2,103}$=3.01). AgoNoBodyC increased from Week1 to 2 and stayed on this level until Week4 (main effect period: P<0.001, $F_{2,105}$=13,53), independent of horn status. The number of injuries (mostly superficial lesions and scars) increased in the course of the experiment. Total number of injuries in the 28 hornless goats were 28, 30, 45 and 57, respectively, for the four periods and 7, 10, 23 and 39, respectively, in the 27 horned goats. Most injuries in hornless goats occurred at the front region, where the horns would grow. Results show, that regrouping should be avoided in goats. While physiological stress is highest directly after grouping, the development of a stable social hierarchy takes more time as indicated by the elevated levels of agonistic interactions and injuries in Week4, at least in hornless goats. Yet, having horns may facilitate this development.

Social behaviour in groups of intact organic fattening boars – a cause for concern?

Jeannette C. Lange and Ute Knierim
University of Kassel, Farm Animal Behaviour and Husbandry Section, Nordbahnhofstr. 1a, 37213 Witzenhausen, Germany; jlange@uni-kassel.de

While it is suggested to avoid castration of male pigs for animal welfare reasons, it should be recognised that other welfare problems may arise due to the boars' social behaviour. Under certain housing conditions, increased frequencies of lameness as well as skin and penis injuries have been reported. The aim of this study was to investigate this issue under commercial organic conditions in Germany. We compared 41 boar and 32 control groups (26× barrows, 3× mixed, 3× gilts) on five farms. At three observation points (around 80 kg live-weight, at slaughter age before and after marketing of first pigs), agonistic and mounting behaviour were observed by continuous behaviour sampling for 15 minutes every alternate light hour from 48 h videos (same number of light hours per farm and batch in boars and controls). Concurrently all 625 boars, 433 barrows and 83 females of these groups were scored for skin lesions and lameness according to the Welfare Quality® protocol. Randomly selected penises of 123 boars and 14 barrows were scored for lesions. Prevalences of pigs being lame at least once in the three scorings were analysed by Chi-square-test. Frequencies of interactions and lesions (group means) were analysed with mixed models in R, with the random factors 'group', nested in 'batch' and within 'farm'. Fixed factors were 'boars vs controls' and, if significant, 'season' and either 'age' or 'observation point'. Further factors (including group size and stocking density) were not significant. More biting and head knocking (2.1 events vs 1.5/pig×hour, $P<0.001$), more fighting (0.4 events vs 0.1/pig×hour, $P<0.001$), and more mounting events (0.44 vs 0.05/ pig×hour, $P<0.001$) in the boar groups did not result in significantly more skin lesions (4.8 vs 4.1 on one body-side, $P=0.48$) or lameness (0.8 vs 2.1%, $P=0.06$). Apparently, the relatively spacious and structured pens allowed a considerable amount of non-injurious social behaviour, which in part may even have been play behaviour. Most lesions were scratches and wounds <2 cm (98.1%). Wounds of 2-5 cm (1.7%) or >5 cm (0.2%) occurred similarly seldom in boars and controls. Also, the maximum lesion number in boars (34) and controls (29) was rather similar. Following the Welfare Quality®-evaluation only 0.1% of controls, but 0.2% of boars were severely injured (WQ-score 2). In addition, 1 control vs 7 boar groups had conspicuously increased lesion levels (>twice than average), with particular contribution to these cases by two farms. All penises of barrows were uninjured whereas in 10% of dissected boars small penile injuries were found. Their absence on two farms with generous straw litter management was noticeable. Results reflect a certain welfare risk due to the higher social activity of intact boars that, however, under organic husbandry conditions and good management can be kept to acceptable limits. In this regard, boar fattening presents a practicable alternative to castration.

Effect of removing the heaviest pigs from a group of grower pigs on aggression and the dominance hierarchy

Sabine Conte[1], Niamh O'Connell[2] and Laura Boyle[3,4]
[1]Agriculture and Agri-Food Canada, Dairy and Swine R&D Centre, Sherbrooke, J1M 0C8 QC, Canada, [2]Queen's University Belfast, Institute for Global Food Security, School of Biological Sciences, Northern Ireland Technology Centre, 97 Lisburn Road, BT9 7BL Belfast, United Kingdom, [3]Institute of Genetics and Animal Breeding, Polish Academy of Sciences, Department of Animal Behaviour and Welfare, ul. Postepu 36A, Jastrzębiec, 05-552 Magdalenka, Poland, [4]Teagasc, Pig Development Department, Animal and Grassland Research and Innovation Centre, Moorepark, Fermoy, Co. Cork, Ireland; laura.boyle@teagasc.ie

Split marketing, whereby the heaviest pigs in a group are sent for slaughter before their penmates, is a common practice. However, there are concerns that every time pigs are removed from a group that there is some re-organisation of the dominance hierarchy (DH) coupled with aggression and that this has negative consequences for pig welfare particularly in entire male production systems. Hence the aim of this study was to investigate the effect of removing the heaviest animals from stable groups of male and female grower pigs on the DH and aggression. Grower rather than slaughter pigs were chosen for this work because slaughter weight pigs can be injured when fighting on fully slatted concrete floors and being heavier, inflict more damage when fighting. At 8 wks of age (17.5 ± 2.0 kg), 64 pigs were assigned to four groups of 8 males and four groups of 8 females. After 5 d all occurrences of aggression and the identity of the pigs involved were recorded over a 25 min period (Pre-test1). Thereafter the troughs were removed from the pen (16:00 h) until 10:00 h the following morning when pigs were subjected to a food competition test (FCT) of 20 min duration. All occurrences of aggressive behaviour in the pen were recorded and a dominance score (DS) was calculated for each pig according to the no. of pigs it displaced and the no. of pigs that displaced it. 12 d after grouping, the 2 heaviest pigs were removed from each pen and the remaining pigs were observed for 25 min (Pre-test2). A FCT was performed the next day as previously described. Data were analysed by ANOVA, Spearmans rank and Pearsons correlation tests and Goodman and Kruskals gamma test in Genstat. There was no difference in the no. of aggressive behaviours performed between the 8 pigs in Pre-test1 and the no. of aggressive behaviours performed by the 6 pigs in Pre-test2 (mean of 4 aggressive behaviours/pig; $P>0.05$). However, there were fewer aggressive behaviours between the 6 pigs of interest (excluding the aggressive behaviour given and received by the two heaviest pigs) at Pre-test1 than at Pre-test2 (i.e. after removal of the two heaviest pigs) (14.8 vs 39.0, respectively, SEM=3.37; $F_{(1,6)}=25.96$; $P<0.01$). There were no gender or time effects on aggressive behaviours performed during the FCT ($P>0.05$). The rank order within the 6 remaining pigs before the removal which was a function of DS measured during the FCT, was not correlated with the rank order after the removal (Gamma=0.19, $P>0.05$). In conclusion, removal of the two heaviest pigs from a pen led to a two fold increase in aggression and re-organisation of the DH. However, it is not known if this represented a permanent re-establishment or temporary confusion of the DH.

Behavior and growth of pigs with divergent social breeding values

Joon-Ki Hong[1], Ki-Hyun Kim[1], Na-Rae Song[1], Hyun-Su Hwang[2], Jae-Kang Lee[2], Tae-Kyung Eom[2] and Shin-Jae Rhim[2]
[1]National Institute of Animal Science, Rural Development Administration, Swine Science Division, 114 Shinbang-1-gil, Cheonan, 31000, Korea, South, [2]Chung-Ang University, School of Bioresource and Bioscience, 4726 Seodongdaero, Ansung, 17546, Korea, South; sjrhim@cau.ac.kr

This study was conducted to characterize the behavior and growth of pigs with divergent social breeding values (SBV). Positive (+, n=5) and negative (-, n=5) SBV groups of finishing pigs (n=70, males: 35, females: 35, age: 64.4 days after birth, weight: 30 kg) were housed in 10 test pens (3.0×3.3 m, 7 pigs/pen). Males and females were evenly distributed between groups. Pigs were observed with the aid of video technology for 9 consecutive hours on days 1, 15, and 30 after mixing. Observation duration and methods are the same for all pens. Moreover, pigs were weighed at approximate 90 kg body weight that was then calculated at the number of days to reach 90 kg to compare the growth between groups. On day 1 after mixing, agonistic behavior was significantly higher in the -SBV than in the +SBV group (Mann-Whitney U test, Z=2.21, P=0.027). Feeding and mate feeding (feeding behavior together with pen-mates) behaviors were significantly higher in the +SBV group on days 1 and 30 after mixing (Z=-5.11~-4.22, P<0.001). Moreover, the day that pigs reached 90 kg was significantly different between the +SBV (128 days) and -SBV (135 days) groups (t-test, t=3.20, P=0.002). The divergent SBV of pigs were affected by responses to social behavior. Such social interactions among pen-mates might affect their growth rate and feed intake. Selection for SBV might be indirect technique for improving the growth performance of pigs. The experimental protocols were reviewed and approved by IACUC, National Institute of Animal Science, Republic of Korea (NIAS 2014-289).

Influence of breed of sire on the occurrence of tail-biting in pigs

Friederike Katharina Warns[1], Felix Austermann[2], Tobias Scholz[3] and Ernst Tholen[4]
[1]University of Goettingen, Department of Animal Science, Albrecht-Thaer-Weg 3, 37075 Goettingen, Germany, [2]Agricultural Chamber of North Rhine-Westphalia, Department of Animal Production, Haus Duesse, 59505 Bad Sassendorf, Germany, [3]Agricultural Chamber of North Rhine-Westphalia, Agricultural Test Center VBZL Haus Duesse, Haus Duesse, 59505 Bad Sassendorf, Germany, [4]University of Bonn, Institute of Animal Science, Endenicher Allee 15, 53115 Bonn, Germany; friederike.warns@agr.uni-goettingen.de

Tail-biting reduces animal welfare in pigs. In this study we focused on genetic effects as a risk factor for tail-biting. 180 non-docked crossbred piglets (Duroc × Topigs20, n=73 [DU]; Piétrain × Topigs20, n=107 [PI]) were weaned at the age of four weeks and divided into two replicates (DG1=28 DU+63 PI; DG2=45 DU+44 PI). According to weight and sire's breed piglets were grouped and raised in four grower pens (16 to 25 pigs per pen with an average of 0,37 m²/pig) in a conventional flatdeck. Water and feed along with chopped alfalfa hay was offered for *ad libitum* consumption. Each piglet's tail was evaluated once a week to determine possible injuries, partial loss, fresh blood and swelling. At the age of ten weeks piglets were transferred to finishing pens. Per replicate, piglets of same sire breed were kept together in one pen. During the finishing period pig's tails were scored 4 (DG1) to 5 times (DG2). Statistical analysis was done with R software (Version 0.99.486) by a GLMM which comprises individual birthweight (BW, covariable), sex (S) and piglets' pen (Bu) as fixed effects, and individual sire × dam of piglet as random factor. Pigs of different breeds were not mixed within the pens, so breed effects were calculated by means of linear contrasts. Six days after weaning tail-biting was observed in one group of DG1 with severe bloody injuries, in other groups only small lesions like scratches were determined. During second half of the flatdeck period and first week in grower pens tail-biting occurred in every group irrespective of R or DG. If tail biting occurred an iodine solution was sprayed to all pigs of the affected pen. There was a significant difference in blood (P=0.024) between breeds and the two replication groups differ significantly in injuries (P=0.017) and swelling (P<0.001). Furthermore there was a highly significant difference for injuries, partial loss and swelling (P<0.001, respectively) between grower pens. The factors injuries, partial loss and swelling showed a comparatively high heritability between 0.217 and 0.279, which was derived from the variance of the random factor sire × dam. Because the results were strongly affected by replication and pen group it is not possible to identify a clear genetic influence on tested evaluation factors. Nevertheless our study indicates a moderate heritability for tail-biting, which motivates to verify the results in further studies.

The effect of illness on social behaviour, cytokine levels and feed intake in undocked boars

Janicke Nordgreen[1], Camilla Munsterhjelm[2], Frida Aae[1], Mari Heinonen[2], Kristin Olstad[3], Torunn Aasmundstad[4], Andrew M. Janczak[1] and Anna Valros[2]
[1]*Faculty of Veterinary Medicine/Norwegian University of Life Science, Production Animal Clinical Science, Ullevålsveien 72, 0033 Oslo, Norway,* [2]*University of Helsinki, Production Animal Medicine, P.O. Box 3, 00014 Helsinki, Finland,* [3]*Faculty of Veterinary Medicine/Norwegian University of Life Science, Companion Animal Clinical Science, Ullevålsveien 72, 0033 Oslo, Norway,* [4]*Topigs Norsvin, Storhamargata 44, 2317 Hamar, Norway; andrew.janczak@nmbu.no*

Tail biting is detrimental to animal welfare and has negative economic consequences. Even though several risk factors are known, they do not reliably predict tail biting outbreaks. Bad health is one of the risk factors for tail biting and the first aim of this study was to test for links between health status and behavior related to tail biting at the individual level. The second aim of this study was to test whether variation in cytokines was related to variation in social behavior. Cytokines, proteins produced by the immune system as a response to infection, but also in response to social stress and contaminated feed, may be mediators between the risk factors of bad health and behaviours increasing the likelihood of tail biting. Cytokines can influence behaviour either in the direction of withdrawal or increased aggression, both of which can increase non-functional social behaviour and thereby the risk of a tail biting outbreak. To investigate this, we collected videos, health data, feeding data and blood samples from undocked boars at a test station farm in Norway. We compared groups with three different diagnoses: osteochondrosis diagnosed by computer tomography scanning (OCSAN; n=30), osteochondrosis diagnosed by clinical examination (OCCLIN; n=14) and respiratory tract disease (RESP; n=13), with healthy controls (CTR; n=37). We tested whether the diagnoses were associated with social behavior and cytokine levels and whether there were correlations between cytokine levels and social behavior. Social behavior differed between categories with OCSCAN pigs receiving more social behavior –both aggressive (ear- and tail biting) and non-aggressive (sniffing)– compared to CTR (P<0.05 for both), and RESP pigs tending to perform more ear- and tail-biting than controls (P<0.1). Ear biting was defined as 'taking the ear of another pig in the mouth followed by an immediate reaction by the receiver'. Tail biting was defined as 'taking the tail of another pig in the mouth followed by an immediate reaction by the receiver'. There were no differences in absolute levels of cytokines between categories, but in particular IL1-ra and IL-12 showed correlations (calculated as Spearman Rank correlations) with several behaviours that have been shown by others to be associated with current or future tail biting activity; IL1-ra levels and tail biting: 0.34, IL1-ra and ear biting: 0.31, IL1-ra and received tail sniffing: 0.30. IL-12 and received tail sniffing: 0.37, IL-12 and received attacks: 0.33. To our knowledge, this is the first published study showing a difference in social behaviour for pigs with different diagnoses and the first to report correlations between cytokines and social behaviour. The possible link between health and tail biting warrants further investigation.

Bash item or clash item – examining pigs' enrichment material priorities and their effects on social behaviour

Linda Wiesner, Hanna Iffland, Rahel Ambiel, Christiane Schalk, Birgit Pfaffinger and Volker Stefanski
University of Hohenheim, Behavioral Physiology of Livestock, Garbenstr. 17, 70599 Stuttgart, Germany; wiesner.linda@gmail.com

Public opinion on the question of how farm animals should be treated has changed remarkably over the last decades. Enhanced housing to improve animal welfare of livestock is one important demand by many consumers in the EU. The question is, however, how welfare can be determined. In the past, indicators for poor welfare, such as injuries or bad physical condition, were used to assess the welfare status of animals. The evaluation of welfare and positive affective state of animals is still a challenge. One approach to assess 'what animals want' is the use of preference tests. Preference testing is a suitable method to draw conclusions regarding the importance of key housing elements. The aim of the present study was to examine pigs' preferences for different enrichments and the enrichments' effects on social interaction among the pen mates. The data are the first of a series of experiments in a larger animal welfare project. 12 castrated male pigs (German Landrace × Piétrain) were housed in six pairs of two. They were about 3 months of age and weighed about 40 kg each at the beginning of the experiment. In a successive preference test, animals were offered access to 4 pre-selected enrichments (soil, straw, metal chains and sisal ropes) consecutively. Each enrichment was presented for 4 days, provided that the order of enrichments was determined randomly. The pigs' behaviour was observed by video recordings on one control day prior to enrichment (empty pen), as well as on the first and third day of enrichment access. Material manipulation and social interactions were analysed by 15 intervals of 6 minutes (1.5 h observation time per day) by continuous recording distributed throughout daylight period (9:00-18:00 h) with The Observer XT©. Preliminary analysis indicates that on the first day of enrichment pigs frequented the enrichment materials soil and straw on average 48 times and 42 times, while the enrichment objects sisal rope and metal chain were used on average 18 and 15 times per 1.5 h daily observation time, respectively. So the pigs frequented enrichment materials soil and straw significantly more often than enrichment objects metal chain and sisal rope ($P<0.003$), whereas there was no significant difference among enrichment materials as well as among enrichment objects ($P>0.05$). Analysis of social data before and after enrichment is in progress and results will be presented at the meeting. So far, we can conclude that our preference test is a useful tool to examine pigs' priorities for enrichments. Following experiments will also include the measurement of physiological parameters (e.g. urinary cortisol) to validate the importance of certain enrichments, especially if access is subsequently denied. We are confident that this approach will allow us to decide which enrichments are of particular importance for pigs and therefore could contribute to the improvement of housing conditions.

Effects of long-term selection for productive traits on sexual behaviour of Taiwan Country chicken

Hsi-Wen Hung
Hualien Animal Propagation Station, Livestock Research Institute, No.38, Sec. 6, Ji'an Rd., Ji'an Township, Hualien County, 973, Taiwan; gingsu@gmail.com

Taiwan Country chickens are vigorous meat-type chickens which is one of the most important chicken breeds in Taiwan. Our hypothesis is that long-term selection of meat and egg production will not only influence the breed on the genetic level but also on the behavioral level. Chickens selected based on their resource allocation of meat and egg production will have different behaviors, and chickens of non-selected strains will retain their own resource allocation percentage and the corresponding behaviors. This study compared the sexual behavior of two long-term selected strains, B for meat production and L2 for egg production, and two randomly-bred local breeds, Quemoy and Shek-ki. Paired mating tests were conducted at 25-26 and 35-36 weeks of age. Each male was paired with a different female of the same strain for four different days. During the test, male and female were separated by a curtain for 20 min to adapt to the environment. After removing the curtain, chickens' behaviors were recorded for 5 minutes. The result shows B strain females performed less crouch in both age period than any other strains ($P<0.05$). The frequency of crouch response in females was over 47% at 35-36 weeks of age, except B strain. Quemoy males performed more waltz, attempt-to-mount, mount and successful mating in both age period than any other strains. The selected strains had less successful mating than did the non-selected strains. Each strain had higher successful mating at 35-36 than at 25-26 weeks of age. In all strains, over 50% males displayed force mating and no significant difference among four strains. Except for B strain, the frequency of force mating decreased with age. In conclusion, long-term selection for productive traits might influence fowls' mating behavior, and fowls will improve mating behavior by age and experience.

Key-note:
What can facial expression reveal about animal welfare? Supporting evidence and potential pitfalls.
Kris Descovich[1,2,3], Jennifer Wathan[4], Matthew Leach[5], Hannah Buchanan-Smith[3], Paul Flecknell[6], David Farningham[7] and Sarah-Jane Vick[3]
[1]*University of Queensland, Centre for Animal Welfare and Ethics, 'White House' Building 8143, Gatton 4343, Queensland, Australia,* [2]*Unitec Institute of Technology, Environmental and Animal Sciences, Private Bag 92025 Victoria Street West, Auckland 1142, New Zealand,* [3]*University of Stirling, Psychology, Faculty of Natural Sciences, Stirling FK9 4LA, United Kingdom,* [4]*University of Sussex, School of Psychology, Room 2A13 Pevensey Building, Falmer BN1 9QH, United Kingdom,* [5]*Newcastle University, School of Agriculture, Food and Rural Development, Agriculture Building, Newcastle upon Tyne NE1 7RU, United Kingdom,* [6]*Newcastle University, Comparative Biology Centre, Newcastle upon Tyne, NE1 7RU, United Kingdom,* [7]*Medical Research Council, Centre for Macaques, Porton Down, Salisbury SP4 0JQ, United Kingdom; kdescovich@unitec.ac.nz*

Animal welfare is a key issue for all industries that use or impact upon animals. The assessment and improvement of animal welfare is dependent on measures that are both reliable and valid and these may include behavioural, physiological, and clinical indicators. Animals may respond to internal and external environments both in generalised (e.g. species-specific) and individual ways. Although no single measure can provide a comprehensive picture of animal well-being, attention in animal welfare science is increasingly turning towards the determination of affective states. Affective states are drivers of good or poor welfare, but are challenging to measure in non-human species, because verbal self-report is not possible. Facial expressions are extensively used as measures of emotion in humans, however they are infrequently used in animal studies, despite evidence that facial changes indicate pain states in many mammals. This empirical review of the science aims to assess whether the measurement of facial expressions is valuable within an animal welfare context. Evidence for facial representations of underlying affective states in animals will be discussed, as well as the relevance of facial communicative signalling to welfare. Facial expressions in mammals are widespread and many movements are conserved across species. Evidence suggests that facial expressions have the potential to indicate psychological and emotional experiences in animals, and can provide information on temporal or stimuli specific reactions. Facial expressions also have social, reproductive and health functions and can therefore be more broadly relevant to welfare assessment than exclusively as indicators of affective state. It is proposed that irrespective of the proximate causes of facial expressions, which are primarily communication and affective state, facial muscle movements or changes in the frequency of such, may be reliably linked to underlying welfare states. Validated tools for measuring facial movement are outlined, and the potential of expressions as honest signals are discussed, alongside other challenges and limitations to facial expression measurement within the context of animal welfare. It is concluded that the measurement of facial behaviour in animals is a useful but infrequently used measure that complements existing tools in the assessment of welfare.

Do dogs smile when happy – an objective and comparative study of dog and human facial actions in response to emotional trigger

Cátia Correia-Caeiro, Kun Guo and Daniel Mills
University of Lincoln, Brayford Pool, Ln58LP, Lincoln, United Kingdom; ccorreiacaeiro@lincoln.ac.uk

One way of investigating emotion cognition processes in animals is by applying comparative psychology methods and look at phylogenetically distant species facing similar present adaptation problems (e.g. domestic dogs-humans). The dog is unique in having shared the same social environment with humans for more than 10,000 years, facing similar challenges and acquiring a set of socio-cognitive skills in the process. Furthermore, the orofacial musculature seems to be relatively well conserved across mammal species, including the domestic dog. Given this commonality in facial musculature, and its known link to emotion expression in humans, it might be expected that similar emotional reactions share communicative correlates in other animals as well. Therefore, this hypothesis was tested. The spontaneous and naturalistic facial behaviour of 50 humans and 100 dogs was analysed and compared, by coding available database videos (e.g. AM-FED, Youtube). These videos were selected based on a list of criteria ranging from quality of image to emotional trigger presence. No concerns regarding animal welfare were raised, since already available videos recorded in day-to-day situations were used. Furthermore, the ethical committee in the School of Psychology, University of Lincoln approved this study. The objective and anatomically-based coding tools, Facial Action Coding System (FACS) and DogFACS, were applied in a range of contexts associated with four classes of emotional response: a) happiness, b) positive anticipation, c) fear, and d) frustration (e.g. thunderstorm-phobic dog, human winning a high stake game). The full range of facial movements in both species was coded, including Action Units (AU), Action Descriptors (ADs) and Ear Action Descriptors (EADs); these are the basic units of independent muscle movement that form facial expressions. Significant differences were found between each emotional response: fearful dogs displayed more AD126 (panting); dogs anticipating a positive event, showed more AD37 (lip wipe), AD137 (nose lick) and EAD102 (ears adductor); and happy dogs, produced more AU27 (mouth stretch); however, frustrated dogs did not display higher rates for any of the facial actions. There were also significant differences between species: when fearful, dogs displayed more AD19 (tongue show) than humans; when frustrated, dogs produced more AU45 (blink); when anticipating a positive event, dogs displayed more AD19 and AD37; and when happy, dogs showed more AD19. This study reports two new important scientific findings: Dogs, like humans, produced differentiated facial movements in response to a particular emotional context; However, the individual facial actions were different between humans and dogs, and, consequently, different facial expressions of emotion were displayed in response to comparable emotional triggers. As such, it challenges some of the arguments relating to the shared mammalian origin of emotion expression, as similar muscle groups are being used to produce different facial expressions of emotion in dogs and humans. These results have important theoretical and practical implications not only for the origin of emotion expression, but also for how humans interact with dogs.

Facial expression of harbour seal pups (*Phoca vitulina*) in response to painful stimuli
Amelia Mari Macrae, David Fraser and Joanna Makowska
University of British Columbia, Animal Welfare Program, 2357 Main Mall, Vancouver, V6T 1Z4,
Canada; amarimacrae@gmail.com

Facial expressions have proven to be a repeatable, accurate and valid way to identify pain in multiple species. Currently, there are no clear species-specific indicators of pain in harbour seals (*Phoca vitulina*) and facial expression has not been examined in any species of pinniped. The aim of this study is to investigate if changes in harbour seals' facial expressions change immediately after a painful procedure. Nineteen pups (healthy, >60 d old) admitted to a Canadian rehabilitation facility had their facial expressions video-recorded during routine procedures of tagging (tag placed inter-digitally through axillary webbing of hind flipper) and micro-chipping (subcutaneous). Analgesia is not standard practice for these procedures. Pups were filmed for 6 min: 2 min before, 2 min during tagging and chipping and 2 min after. For each seal, a 90-sec segment of both before and after the procedure was extracted (within 5 s of noxious stimuli) from the videos. Two observers, blind to treatment, watched the 38 video clips in randomized order. Orbital tightening, bunching of the nose, whisker position, mouth position, as well as behaviours such as vocalizing, looking around (inquisitive), struggling, trembling, open-mouth breathing and biting were scored as present or absent for each 1-sec interval of each 90-sec clip. Observers also assigned an overall pain score (pain present or absent, scored subjectively) for the entire clip. Observers were considered to be 'experienced' (>8 yr working with pinnipeds) and 'inexperienced' (no direct experience with pinnipeds). Paired sample t-tests showed that orbital tightening increased from before to after the procedure (t=10.80, df=18, P<0.0001), whereas the behaviours of looking around (t=-5.92, df=18, P<0.0001), struggling (t=-3.43, df=18, P<0.003) and trembling (t=-2.91, df=18, P<0.01) all decreased from before to after the procedure. Inter-observer reliability ranged from r=0.89 to 0.92 for the four types of behaviour. For both observers, the subjective scoring of whether pain was present or absent was very accurate (95% for experienced and 89% for inexperienced respectively). Each animal acted as its own control; however, a separate trial with 5 pups had shown no change in the behavioural responses to sham tag/chip procedures indicating the response seen in this trial was due to pain rather than handling. As there are currently no published pain scales based on facial expression of this species, these results show promise for facial expression and other behaviour to be used as a tool for acute pain assessment in seals.

The startle response as a measure of fear in dairy calves

Sarah J.J. Adcock and Cassandra B. Tucker
Center for Animal Welfare, Animal Science, University of California, Davis, CA, 95616, USA;
sadcock@ucdavis.edu

There are, at present, no well-validated tests for fear in cattle. A particularly powerful tool that has been used to study fear in humans and rodents is the startle response – a ubiquitous, cross-species defensive reflex to sudden stimulation. Our objective was to assess the validity of the startle response as a measure of fear, defined here as the behavioral and physiological response to a perceived threat, in dairy calves. We predicted that calves exposed to predator odor, a fear-eliciting stimulus, would have a higher startle magnitude, elevated heart rate, and longer latency to feed in response to a sudden noise than calves exposed to the odor of a herbivore or water. Thirty-four calves were assigned to 1 of 3 treatments (coyote urine, deer urine, water) in a between-subjects design. Each calf was fitted with an accelerometer and heart rate monitor and tested individually in 10 min trials in a familiar arena on 3 consecutive days. A bucket feeder containing 500 ml of milk replacer was secured to the side opposite the arena entrance. Immediately before the trial, the odor was dispersed into the arena with fans. The startle stimulus, a 0.4 s, 105±2 dB burst of white-noise, was delivered upon the calf's arrival at the feeder. We assessed treatment (coyote, deer, water) and test day (1, 2, 3) effects on startle magnitude and heart rate with linear mixed models using the restricted maximum-likelihood method and calf identity fitted as a random effect. The effect of treatment and test day on latency was modeled using a parametric survival model. Correlations between variables were made using Spearman's rho. The magnitude of the behavioral startle response was higher in the coyote treatment than in both controls on the first day of testing (mean total acceleration on day 1 ± SE: Coyote: 199±15 g; Deer: 166±3 g; Water: 172±8 g; Treatment×Day: $F_{4,50}$=2.56, P=0.050). Calves exposed to coyote odor tended to take longer to return to the feeder following the sudden noise compared to control animals (mean ± SE: Coyote: 11±5 s; Deer: 4±2 s; Water: 4±1 s; Treatment: X^2_2=8.32; P=0.016; post-hoc tests: P≤0.086). The cardiac startle response did not differ between treatments (mean heart rate ± SE: Coyote: 164±6 bpm; Deer: 167±6 bpm; Water: 177±6 bpm; $F_{2,30}$=1.29, P=0.290) nor days ($F_{2,52}$=0.02, P=0.979). Startle magnitude and latency to return to the feeder were positively correlated (rho=0.60, P<0.0001). Unexpectedly, heart rate correlated negatively with both startle magnitude (rho=-0.26, P=0.019) and latency to return to the feeder (rho=-0.46, P<0.0001), which suggests fear brachycardia – a defense mechanism characterized by a decrease in heart rate – may accompany the startle response. In conclusion, the results provide support for using the behavioral startle response as a measure of fear in dairy calves.

Pre-weaning environment affects pigs' emotional reactivity

Oceane Schmitt[1,2], Laura Boyle[2], Emma M. Baxter[1] and Keelin O'Driscoll[2]
[1]SRUC, Animal Behaviour and Welfare, Animal and Veterinary Science Research Group, West Mains Road, Edinburgh EH9 3JG, United Kingdom, [2]Teagasc, Pig Development Department, Moorepark, Fermoy, Co. Cork, Ireland; oceane.schmitt@teagasc.ie

Artificial rearing systems such as Rescue Decks (RD) are of growing interest to facilitate the rearing of super-numerous piglets born in large litters, where survival is compromised. However, as piglets are separated early from their dam (i.e. 7 days-old), and have less space allowance in the RD than in a farrowing pen, rearing in a RD might affect pigs' emotional reactivity. This study compared the post-weaning reactions of sow-reared (SOW) piglets and artificially reared (RD) piglets (i.e. milk replacer in RD from 7 days-old until weaning) to a variety of novel situations, using four behaviour tests. Tests were carried out on groups in the home pen at approximately 1 week post-weaning (W1; 33±3 days-old, n=26 pens), at transfer to the second stage weaner accommodation (W2; 69±1 days-old, n=18 pens) and at transfer to the finishers accommodation (F; 100±1 days-old, n=14 pens). The tests were: (1) Startle Test (ST): the group's reaction to the opening of a red umbrella in front of the pen (1: >60% froze/startled; 0: no reaction) and the latency to recover from startling (i.e. return to normal activity for 50% of startled group) were recorded; (2) Novel Object Test (NOT): the latency of the first pig to touch a novel object (yellow Frisbee, pink paddle or blue bucket) was recorded; (3) Human-Animal Relationship Test (HART): the reaction of pigs to a familiar human's contact was scored (1: avoidance of stationary human or human contact, 0: acceptance of contact); and (4) Open Door Test (ODT): the latency of the first pig to exit the pen, and the max proportion of pigs outside the pen during 5 min after the pen door was opened were recorded. Latencies were analysed using survival curve analysis. Prevalence of startling reaction and percentage of fearful pigs during HART were analysed using Generalised Linear Mixed Models with the total number of pigs in the pen being a random factor. There was no interaction between time and pre-weaning environment, thus only pre-weaning environment effects are reported. There was no effect of pre-weaning environment on the ST reaction, but SOW pigs took longer to recover compared to RD pigs (16.64±2.16 vs 11.65±1.20 sec; P<0.05). Pre-weaning environment did not affect latency to touch the novel object. However, more SOW pigs showed fear of human contact than RD pigs (0.58±0.05 vs 0.32±0.05; P<0.001). Latency to exit the pen was not affected by pre-weaning environment (P>0.05), but the maximum proportion of pigs observed outside the pen was higher in SOW than in RD (0.83±0.03 vs 0.72±0.03; P<0.05). Overall, RD pigs seemed calmer than SOW pigs in the presence of a human, as indicated by a faster return to normal activity after startling, and more willingness to accept human contact. This could be a result of increased exposure to human contact prior to weaning. However, fewer of these pigs spontaneously left the home pen, indicating greater reluctance to explore a novel environment.

Lying around with nothing to do: boredom in farmed mink

Rebecca K. Meagher[1,2], Dana L.M. Campbell[1] and Georgia J. Mason[1]
[1]University of Guelph, Animal Biosciences, 50 Stone Road East, Guelph, N1G 2W1, Canada,
*[2]University of Reading, School of Agriculture, Policy and Development, Whiteknights, P.O. Box
237, Reading RG6 6AR, United Kingdom; rkmeagher@gmail.com*

Boredom, a negative emotional state resulting from too little or monotonous stimulation, has received little empirical study in non-human animals, but is commonly believed to be prevalent in captive environments. If true, this would mean that millions of animals worldwide are experiencing an aversive state that might lead to other negative consequences such as depression and poor health. Boredom should, by definition, result in motivation to obtain novel stimulation. In a previous experiment, we demonstrated that, compared to mink in elaborately enriched cages previously shown to improve welfare, mink in non-enriched farm cages showed exaggerated interest in a wide range of stimuli, suggesting that they do indeed experience boredom-like states. According to some measures, this also correlated with time spent lying awake when undisturbed. We conducted a second study to attempt to replicate these findings. We used 20 adult male mink, half of which were housed in standard non-enriched cages, while half had been housed in enriched cages since 3 months of age. Spontaneous behaviour (locomotor stereotypies and time spent lying awake but inactive) was assessed by 340 scan samples over 20 days in the winter. Tests of interest in stimulation were conducted in the spring. Mink were exposed to 11 stimuli, categorized as aversive, rewarding or ambiguous based on biological relevance and prior observation, for five minutes each. Interest was measured via the latency to touch, time in contact with, and time oriented to the stimuli, as well as consumption of food treats in three separate tests. MANOVAs showed that time oriented to and in contact with stimuli were higher in non-enriched mink, although the magnitude of these effects differed across stimuli (interaction terms: $F_{9,51}=2.62$, P=0.016 and $F_{7,33}=2.38$, P=0.047, respectively). When analysed separately by stimulus type, for aversive stimuli only, these measures were higher in non-enriched than enriched mink (oriented to: $F_{1,8}=4.12$, P=0.038; in contact: $F_{1,8}=6.97$, P=0.015). Latency to contact and food treat consumption did not differ between treatments in this study. Time lying awake again correlated with average orientation time (general linear model: $F_{1,8}=9.49$, P=0.015) within the enriched treatment. Overall, this study broadly confirms our previous findings that mink in non-enriched cages express a core symptom of boredom.

Can minor, easily applied alterations of routines during the rearing period reduce fearfulness in adult laying hens?

Margrethe Brantsæter[1], Fernanda Machado Tahamtani[2], Janicke Nordgreen[1], Ellen Sandberg[3], Tone Beate Hansen[4], Bas Rodenburg[5], Randi Oppermann Moe[1] and Andrew Michael Janczak[1]
[1]Norwegian University of Life Sciences, Department of Production Animal Clinical Science, Ullevaalsveien 72, 0033 Oslo, Norway, [2]Aarhus University, Department of Animal Science, Blichers Allé 20, DK-8830, Tjele, Denmark, [3]Norwegian University of Life Sciences, Department of Chemistry, Biotechnology and Food Science, Arboretveien 40, 1432 Ås, Norway, [4]Animalia, Norwegian Meat and Poultry Research Centre, Lørenveien 38, 0513, Oslo, Norway, [5]Wageningen University & Research, Behavioural Ecology Group, Adaptation Physiology Group, Droevendaalsesteeg 4, 6708, Wageningen, the Netherlands; margrethe.brantsaeter@nmbu.no

Exaggerated fear-reactions are associated with injuries, smothering, feather pecking and other events that compromise laying hen welfare. Provision of litter during the rearing period may reduce fearfulness. The aim of this study was to test the hypothesis that chicks with access to litter during the first five weeks of life would be less fearful as adults compared to birds reared without access to litter. The hypothesis was tested in 23 commercial aviary layer flocks in Norway. Five rearing farmers divided the pullets into two groups within their rearing houses. During the first five weeks of life, paper substrate, on which food and other particles could accumulate, covered the wire mesh floor in the treatment group, whereas the control group was reared on bare wire mesh. The egg producers were instructed to follow their normal management procedures. At 30 weeks of age, 23 layer flocks (11 control flocks and 12 paper reared flocks) were visited. A stationary person test and a novel object test were conducted to test fearfulness of the adult hens. In addition, data on provision of environmental enrichment was collected as a binary yes/no variable. Provision of environmental enrichment to adult birds tended to reduce the latency to approach within 2 m of the stationary person (P=0.08). For birds without environmental enrichment as adults, access to litter during rearing increased the number of birds that approached the novel object compared with birds reared without paper (P=0.04). For birds with access to environmental enrichment during production, the access to litter during rearing had no effect on the number of birds that approached the novel object as adult (P=0.99). These results indicate that both providing chicks with paper substrate from the first day of life and providing them with environmental enrichment as adults, are practical and simple alterations of management that reduce fearfulness in laying hens.

Assessing anxious states in sheep: a more practical attention bias test

Jessica Monk[1,2], Rebecca E. Doyle[3], Sue Belson[1], Greg Cronin[4] and Caroline Lee[1]
[1]CSIRO, Agriculture and Food, FD McMaster Laboratory Chiswick, Armidale NSW 2350, Australia, [2]University of New England, Natural Resources Building, Armidale NSW 2350, Australia, [3]Animal Welfare Science Centre, University of Melbourne, Alice Hoy Building, Parkville VIC 3010, Australia, [4]University of Sydney, JL Shute Building, Camden NSW 2570, Australia; jessica.monk@csiro.au

Animals experiencing increased states of anxiety show greater attention towards threatening stimuli than calmer individuals, a response termed attention bias. Tests for attention bias potentially offer more rapid assessment of affective state in animals than existing cognitive methods. We have previously developed and pharmacologically validated an attention bias test as a measure of anxious states for sheep. Experiment 1 in the current study aimed to confirm that the responses being measured in the test were directed towards the dog and not just other aspects of the test such window cover movement. Experiment 2 aimed to refine the test, shortening the duration from 3 min per sheep and removing the need to train sheep to feed from a bucket. In each experiment, the attention bias test consisted of an enclosed arena (4×4.2 m) with Lucerne hay in the centre and a small window on one wall. When a sheep entered the test, a dog sitting quietly was visible through the window for 3 s, then the window was covered. The sheep remained in the test for 3 min while behaviours were recorded. Duration of vigilance (head above shoulder height) was analysed using Kruskal-Wallis one-way ANOVA and feeding behaviour was analysed using survival analysis. For experiment 1, sheep either underwent the described procedure or were exposed to an empty window at the beginning of the test instead of a dog (n=20 per group). Sheep exposed to the dog spent more time displaying vigilant behaviour (mean ± s.e. 165±2.8 s vs 147±5.3 s; P=0.004), were less likely to feed (log rank test P<0.001) and were less likely to sniff the window (Fisher's exact test P=0.008) than sheep exposed to an empty window. These findings indicate the behaviours recorded in the test were at least partially a response to the dog. In experiment 2, 60 sheep were divided into 3 groups (n=20) and given either an anxiolytic drug (diazepam, 0.1 mg/kg i.v.), an anxiogenic drug (1-(m-chlorophenyl)piperazine (mCPP), 2 mg/kg i.m.) or a control treatment (saline i.m.) before testing. Behaviours during the first 45 s of testing were also assessed to determine whether the test duration could be shortened. During the first 45 s, mCPP treated sheep spent more time displaying vigilance (35±0.6 s) than control (29±1.6 s) and diazepam (24±2.2 s) groups (P<0.001) and were less likely to feed (log rank test P<0.001). This indicates differences in anxious state may be detected using a shorter test. Only 4 animals in the control and diazepam groups failed to eat the novel feed within 180 s, showing it is not necessary to train sheep prior to testing. Overall the current study refined the attention bias test by removing the need for prior training and shows the test duration can be shortened. It also confirms that the measured responses were towards the dog and provides further pharmacological validation that the attention bias test may be useful for assessing anxious states in sheep.

Development of behavioural tests for WelFur on-farm welfare assessment of foxes

Jaakko Mononen[1], Eeva Ojala[2], Hannu T. Korhonen[1] and Tarja Koistinen[1]
[1]*Natural Resources Institute Finland, Green Technology, Halolantie 31A, 71750 Maaninka, Finland,*
[2]*Kannus Research Farm Luova Ltd., Turkistie 6, 69100 Kannus, Finland; jaakko.mononen@luke.fi*

The WelFur on-farm welfare assessment protocol for foxes includes two behavioural tests. The Feeding Test (FT) is a well validated test that measures human-animal relationship (HAR). It is the only measure in the 'Good HAR' criterion of the WelFur fox protocol. In the test the assessor delivers feed to a hungry fox and records whether the animal eats or not within 30 seconds while the assessor stands near the animal's cage. The Stick Test (ST), in turn, is supposed to measure explorative behaviour, and forms a part of the 'Positive Emotional State' criterion in the protocol. In the test the assessor inserts 20-30 cm of a 150 cm stick into the cage of a fox, and records whether or not the fox touches the stick in an explorative way within 10 seconds. ST is performed as the very first part of collecting data at the cage level, and the assessor tries to stay as far as possible from the cage in order to minimize human effect. ST has not been, however, properly validated, and since exploration may be affected by the presence of human we studied whether the results of FT and ST overlap. In addition we studied whether a Subjective Evaluation of HAR (SE) during the 1.5-3.0 min period when other animal and resource-based data are collected from a cage could substitute for the more laborious FT in the WelFur protocol. In SE the tendency of a fox to approach (low score) or withdraw from (high score) the assessor was scored on a scale from one to five. All adult blue fox (n=354) and silver fox (n=94) vixens on a private Finnish farm were tested with FT, ST and SE during two days in October. The results from FT and ST were compared with Logistic Regression to study the overlap hypothesis. Species was in the model as a blocking factor. SE scores (ordinal scale) between the animals eating and not eating in FT were compared with Mann-Whitney U-test separately for the two species to study the substitution hypothesis. Logistic Regression revealed that FT and ST were not independent of each other (P<0.01), but species had no effect (P=0.65). The 93% precision (251/271 foxes; species pooled) of ST in predicting FT indicates that ST measures to a great extent the same character as FT, most probably HAR. On the other hand, sensitivity of ST in predicting FT was only 69% (251/362 foxes; species pooled), i.e. ST failed to recognize almost one-third of the foxes eating in FT. The SE score was lower (P<0.01, Mann-Whitney) in foxes eating than not eating in FT in both species: blue foxes, 2.5-3.0-3.5 (quartiles Q1-Q2-Q3, n=282) vs 3-3-3 (n=72); silver foxes, 1.0-2.0-2.5 (n=81) vs 2.5-3.0-3.5 (n=13). However, the distributions of the scores overlapped markedly between the two groups of foxes in both species. We conclude that ST is rather a HAR test than a test of explorative behaviour, and ST could probably substitute for FT in the WelFur fox protocol. Instead, SE would require further development aiming at increasing its discriminative power before it could replace FT.

Dietary tryptophan supplementation and affective state in pigs

Jenny Stracke[1,2], Winfried Otten[2], Armin Tuchscherer[3], Maren Witthahn[2,4], Cornelia C. Metges[5], Birger Puppe[2,4] and Sandra Düpjan[2]

[1]University of Veterinary Medicine Hannover, Institute for Animal Hygiene, Animal Welfare and Farm Animal Behaviour, Bischofsholer Damm 15, 30173 Hannover, Germany, [2]Leibniz Institute for Farm Animal Biology (FBN), Institute of Behavioural Physiology, Wilhelm-Stahl-Allee 2, 18196 Dummerstorf, Germany, [3]Leibniz Institute for Farm Animal Biology (FBN), Institute of Genetics and Biometry, Wilhelm-Stahl-Allee 2, 18196 Dummerstorf, Germany, [4]Faculty of Agricultural and Environmental Sciences, University of Rostock, Behavioural Sciences, Justus-von-Liebig-Weg 6, 18059 Rostock, Germany, [5]Leibniz Institute for Farm Animal Biology (FBN), Institute of Nutritional Physiology, Wilhelm-Stahl-Allee 2, 18196 Dummerstorf, Germany; jenny.stracke@tiho-hannover.de

The assessment and provision of welfare in farm animals has become a major issue in animal science. Affective states in general, and positive affective states in this particular study, are a key element for providing good welfare. As the serotonergic system plays a central role in regulating affective behaviour, an increase in centrally available serotonin (5-HT) via dietary supplementation of its precursor tryptophan (TRP) might be an approach to induce positive affective states. Therefore, the aim of our study was to investigate the effects of dietary TRP supplementation on brain TRP metabolism and 5-HT levels, but also on affective state (cognitive bias test) and behavioural reactivity (combined open field/novel object test) in pigs. The study was conducted on 52 female, juvenile pigs in total (20 for part 1; 32 for part 2). All experimental procedures were ethically approved (LALLF M-V/ AZ:7221.3-1-066/13). Animals were fed a standard diet until eight weeks of age, then feed was changed for all animals, with half the animals receiving a diet with a standard TRP content (2.5 g/kg; control), while the other half received a TRP enriched diet (10.2 g/kg; TRP+). In part 1 of our study, we investigated the effects of the dietary TRP supplementation on TRP metabolism in brain areas related to affective and cognitive processing (data analysed blind to treatment). We found significantly increased concentrations of TRP (all $F>82.7$; all $DF=1,15$; all $P<0.001$) and its metabolites in nearly all analysed brain tissues. In part 2 of our study, we analysed the effects of these alterations on the affective state as measured in a cognitive bias test, namely the spatial judgement task (SJT) which we had validated for pigs in a previous study (data not shown). Here, animals had to learn to discriminate between a positively reinforced and a negatively reinforced location, and were subsequently confronted with ambiguous, intermediate locations. Additionally, a combined open field/novel object test (OFNO) was conducted to measure behavioural reactivity (activity/exploratory behaviour). The TRP enrichment revealed no significant behavioural changes in the OFNO tests (all $P>0.05$). In the SJT, the TRP+-group showed more pessimistic behaviour after dietary change than before. Therefore, our results do not support the suggestion that TRP supplementation induces positive affective states and thus improves animal welfare in pigs.

Playing or exploratory behaviour in pigs: neurohormonal patterns in mini-pigs and suggestions for welfare management

Míriam Marcet Rius, Alessandro Cozzi, Cécile Bienboire-Frosini, Eva Teruel, Camille Chabaud, Philippe Monneret, Julien Leclercq, Céline Lafont-Lecuelle and Patrick Pageat
Research Institute in Semiochemistry and Applied Ethology, Quartier Salignan, Apt 84400, France; m.marcet@group-irsea.com

The European Directive 2008/120/EC laying down minimum standards for the protection of pigs establishes that pigs must have permanent access to a sufficient quantity of material, such as straw, to enable proper investigation and manipulation activities. The aim of this study was to investigate benefits that different types of enrichment, triggering different behaviours, like exploratory with straw or playing with dog toys, could provide in pigs. Peripheral oxytocin and its evolution over time was measured. Play behaviour is a commonly observed and characteristic behaviour of young mammals, and exploratory behaviour, a high priority behaviour in pigs, *i.e.* one of the behavioural needs. Housing and experimental procedure were carried out according to European legislation and in the respect of ethical principles. It was approved by IRSEA's Ethics Committee (125) and the French Ministry of Research (AFCE_201602_01). 18 minipigs were divided into 2 halls and housed in pens of 2 individuals. 'Enriched group' received dog toys during play sessions, every day during a total of 30-40 minutes, for 3 weeks, and after it, straw was provided continuously in the pens for 3 more weeks, whereas other group, 'Control group' did not have any material during all this period. Animals of 'Enriched group' were subjected to 10 minutes' video recording per day, in their pens, to analyse the behaviour during the interaction with each material. One blood sample had been drawn before play session and before the provision of straw (T_0), and another one 10 minutes after playing or exploring during 10 minutes (T_1). This procedure was the same for each animal during the 6 weeks, including the animals of the 'Control group' that did not explore nor play. The mostly found behaviours during the interaction with toys were: object play, described as biting, sniffing, pushing, kicking, licking, and chewing the toy, as well as tail movement behaviour, considered as a possible indicator of positive emotions. The mostly found behaviours during the interaction with straw were exploratory behaviour, especially rooting, as well as tail movement behaviour. For play behaviour, oxytocin was significantly higher at T1 than at T0 in the 'Control group' (T0: 5.00±2.65 vs T1: 9.22±11.56; P=0.00); in 'Enriched group' there was no significant difference between T0 and T1 (T0: 4.98±2.97 vs T1: 5.37±3.47; P=0.87). Regarding exploratory behaviour, oxytocin was significantly higher at the 3rd week than the 2nd week (2^{nd}w: 3.90±2.42 vs 3rd w: 6.16±2.72; P=0.02). Results suggest both types of material produce positive emotions and allow the development of natural behaviours. As the provision of toys could be an easier measure for the actual system than the provision of straw, we recommend it as a measure to improve welfare.

Low rate of stereotypies in pregnant sows is associated to fear related behaviours in their offspring

Patricia Tatemoto, Thiago Bernardino, Luana Alves, Marisol Parada Sarmiento, Anna Cristina De Oliveira Souza and Adroaldo José Zanella
Center for Comparative Studies in Sustainability, Health and Welfare, Department of Veterinary Medicine and Animal Health, USP – VPS, Avenida Duque de Caxias Norte, 225, Pirassununga, SP, 13635-900, Brazil; patricia.tatemoto@usp.br

The commercial sow housing systems during gestation are frequently environments considered barren by their low complexity. The reduction of environmental stimuli has been considered a stressor and cause of stereotypies. The effects of prenatal stress in the brain can generate changes in the emotionality in the offspring. We studied the effects of the expression of stereotypies performed by the mother on the welfare indicators in piglets. We evaluated sows kept in group housing pens (n=11; low rate abnormal behavior = 4, high rate abnormal behavior = 7), recording the behaviors on gestation days 88, 89, 91, 92, 106 and 107, for two consecutive minutes, with three replicates, totalizing six minutes per female in each day. In the first time, we observed 35 sows maintained in barren pens or pens with straw (hay). To this part of analysis, we considered just the females kept in the barren pen (control group) with expression of abnormal behaviors which was consistent throughout the days. Thus, sows with low rate abnormal behavior were those that never presented the behavior and females characterized by a high rate of abnormal behavior were those that presented stereotypies for at least 50% of the days of observation. To access emotionality, we subjected the piglets (one couple per sow) individually to an open field test and to a novel object test (totalizing 10 minutes), we recorded behaviors such as latency for activity, activity duration, exploratory behaviors, freezing and vocalization. In the open field test, we showed that piglets from females with low rate abnormal behavior rate presented higher latency for activity (Mann-Whitney, P=0.04) and less activity duration (walking through the pen; P=0.01). To measure aggressiveness, we counted the number of skin lesions in each piglet using photographs, on 28, 29, 35, 36, 42 and 43 days after weaning. There was only difference throughout time (Anova, repeated measures; P=0.0001), where the lesions were reduced as the hierarchy stabilized within the group. We conclude that the rate of abnormal behavior expressed by the mother is related to more fearful offspring. These changes in the phenotype of offspring indicate changes in the limbic system associated with fear.

Influence of novelty and housing environment on play and tail wagging behavior in pigs

Lisa Christine McKenna and Martina Gerken
University og Goettingen, Animal Sciences, Albrecht-Thaer Weg 3, 37075 Göttingen, Germany;
lisa.mckenna@agr.uni-goettingen.de

To improve the welfare of pigs in intensive husbandry systems, positive emotions should be stimulated. Indicators of positive emotions in pigs are e.g. tail wagging and play behaviour. The present study was conducted to investigate whether female pigs kept in housing environments with different enrichment levels show differences in the occurrence of play and tail wagging behaviour when confronted with novel objects. A total of 36 female animals were tested over a period of 2 weeks when the animals were between 5 and 7 weeks of age. The experimental animals were kept either in a non-enriched (NE) stable (barren, except for a thin layer of sawdust), an enriched (E) stable (sawdust and straw bedding, 2 m^2/animal) or in a super-enriched (SE) stable (sawdust and straw bedding, 2 m^2/animal plus additional enrichment which was changed weekly). In every environment, there was one test group (n=6) and one control group (n=6). For behavioural testing, the animals were led into the test room in groups of three to avoid social isolation stress. There, three novel objects (the same for each animal) were laid out for the animals of the test groups to explore. For the control animals, the test room remained empty. The novel objects consisted of: a rubber duck, a rubber dog toy filled with grapes, a wallow and potting soil. Test objects were chosen based on their potential to trigger positive emotions in the animals. The animals were videotaped during the testing period of 7 minutes. Play and tail wagging behaviour were recorded continuously. Each animal was tested 4 times (twice per object). The frequency of occurrence of these behaviours was added up and averaged for the test and control groups. This calculation resulted in a Positive Emotion Score (PES) which was compared between test and control groups and the different housing environments. Residuals of this variable were not normally distributed and were therefore analysed with the Mann Whitney U test. We have found that the test groups showed significantly higher PES than the control groups (P=0.001). We also found a tendency for higher PES in the test animals of the NE compared to the E and SE environment (P=0.74, P=0.22, respectively) (PES NE: 4.27 and 0.46, E: 2.25 and 0.10, SE: 1.73 and 0.79 for test and control animals, respectively). Previous studies showed rebound effects in pigs housed in barren environments. The similar tendency found in NE animals was unexpected, as these animals were provided with sawdust. We suggest that novel objects potentially cause positive emotions in pigs as measured by play and tail wagging behaviour.

The effect of different stimuli on the 'stick test' should be taken into account when measuring the mink's temperament

Britt I.F. Henriksen, Jens Malmkvist and Steen H. Møller
Department of Animal Science, Blichers Allé 20, 8830 Tjele, Denmark; britt.henriksen@anis.au.dk

The temperament of mink can be measured in a voluntary approach-avoidance test called the 'stick test'. The test is measuring the mink's reaction when a wooden tongue spatula is put into the front of their cage through the wire netting. Their reaction is categorised as 'fearful', 'exploratory', 'aggressive' or 'undecided'. We aimed to test the effect of different stimuli on the temperament score obtained in a standardised 'stick test' of mink. We tested the influence of (1) additional cage shelf in the front part of the cage (vs none), (2) distance from test person to mink, (3) familiar vs unfamiliar dressed test person, and (4) two types of wooden test spatula (small vs large). We hypothesise that (1) mink may use the cage shelf as a safe place, thus increasing their exploratory behaviour in the test, (2) fear increases when the test person is bending over the cage, (3) unfamiliarity in colour and smell of test person's coverall results in decreased exploratory behaviour and increased distance to stimuli and that (4) the type of wooden spatula does not affect the score. We tested 600 pair-housed (male/female) brown juvenile mink in October/November 2015 and 500 brown female mink (first and second parity) in January/February 2017. In 2015 the mink were divided into two groups of 150 cages where one group was tested with or without the human tester bending over the cage, and the other group was tested with or without the animals having access to an extra shelf in the front part of the cage. In 2017 the mink were divided into five groups of 100 cages per group testing combinations of the human tester having white or dark blue disposable coverall or a dark blue coverall from the farm, tongue spatula (150×17 mm) or a coffee stick (140×5 mm). Each animal was tested twice with a different combination. Statistical analysis (prevalence of scores) was done in logistic binomial mixed models taking repeated measures and two mink per cage into account. We did not find any significant effect of the human position on the scores ($P>0.05$). However, of the non-exploratory animals, the human tester bending over the cage resulted in fewer animals in the first half of the cage close to the test stimuli ($\chi^2=4.0$, $P<0.05$). The prevalence of exploratory animals was significantly higher ($\chi^2=29.1$, $P<0.0001$) and the prevalence of fearful and undecided animals significantly lower (fearful: $\chi^2=28.8$, $P<0.0001$ and undecided: $\chi^2=7.3$, $P<0.001$) when the animals had access to an extra shelf. Thus, mink lying on the shelf during testing were more often scored explorative ($\chi^2=41.6$, $P<0.0001$). This is in line with other studies showing that a shelf increase exploration/decrease flight in the stick test of mink. Data from the trials in 2017 is upcoming. We conclude that the position of the test person has minor influence on exploration/fear in mink, not affecting the score, whereas a cage shelf close to the test area significantly increases mink scored explorative.

Stress in male Japanese macaques living in vegetated and non-vegetated enclosures

Josue S. Alejandro, Michael A. Huffman and Fred Bercovitch
Primate Research Institute, Kyoto University, Ecology and Social Behavior, 41-2, Kanrin, Inuyama
City, Aichi Prefecture, 484-8506 Japan, Japan; josue.pastrana.64s@st.kyoto-u.ac.jp

Improving captive environments for primates has been an important tool to enhance animal welfare. To investigate the benefits in which living in naturalistic environments decreases stress and promotes general animal well-being we observed two outdoor housed groups of Japanese macaques (*Macaca fuscata*) in the Primate Research Institute (PRI). The two groups are located in the same area (similar proximity to road, human contact) and were selected for the study based on similar age-sex class composition (5 mature males, 15-20 mature females, and immatures for both conditions), group size (42-45 individuals), rearing histories, and year of establishment. The vegetated enclosure (3,900 m^2) contains trees, a pond, natural ground cover as well as artificial perching. The enclosure is divided in two sections and animals are moved once a year to allow for re-growth of the vegetation. The non-vegetated enclosure (960 m^2) is a concrete outdoor enclosure with perching, swings, climbing structures, artificial river, and platforms with shelters to protect animals from the weather. 10 individuals (5 mature and 5 immature) from both conditions were observed and their behaviors recorded using 10 minute continuous focal sampling across seasons. All observations were conducted inside the enclosure and the observer habituated to both groups for a period of 3 months. Their activity budgets, rates of agonistic, affiliative (groom, play) and abnormal behaviors, as well as coat conditions were recorded using a 0 to 1 scale. We found that animals in the naturalistic enclosure had activity budgets more similar to their wild counterparts; they spent significantly less time moving ($U=17$, $n_1=10$, $n_2=10$, $P=0.013$), less time in agonistic interactions during the mating season ($T=4.624$, $df=19$, $P=0.001$), and immatures spent more time in social play than animals in the non-vegetated enclosure ($U=2$, $n_1=10$, $n_2=10$, $P=0.012$). We found no differences in social grooming or self-grooming, but we did note significantly better coat conditions in mature individuals in the vegetated enclosure ($U=0$, $n_1=5$, $n_2=5$, $P=0.008$) and a higher incidence of abnormal behaviors in the non-vegetated enclosure. Animals in highly enriched, vegetated enclosures allocated their time across activity patterns in a proportion more similar to that of their wild counterparts. Behaviors associated with optimal living standards, such as playing and foraging have an important role in reducing stress and they seem to be more prevalent in highly enriched enclosures. Perhaps because living in an enriching environment is less stressful, animals spend less time in aggression and stereotypic behaviors than animals living in conventional non-naturalistic environments. One limiting factor in our study was the difference in size of the enclosures, which might have influenced on rates of aggression. Nonetheless, monkeys living in enclosures that promote species-specific behavior seem to prevent or at least ameliorate welfare concerns. Therefore, to the extent possible, primates in non-vegetated environments should have their condition improved in order to provide a healthier living area.

Algorithm for real-time classification of mental responses of horses during physical activity

Vasileios Exadaktylos[1], Anne Marleen Van Aggelen[2], Tomas Norton[2] and Daniel Berckmans[2]
[1]BioRICS, Technologielaan 3, 3001 Leuven, Belgium, [2]KU Leuven, Biosystems Engineering, Kasteelpark Arenberg 30, 3001 Leuven, Belgium; marleen.vanaggelen@kuleuven.be

A horse is a living organism and consequently each horse has an individual response to a specific situation. The physical and mental status of a horse plays an important role in the animal's behaviour. In order to understand the behaviour of horses better, it's important to gain insight in mental status beside the physical status of the animal. Numerous ways exist to quantify the physical status of a horse (e.g. medical examination, exercises tests, etc.) by audio-visual scoring in a manual way. Although this methodology exists, the quality of the mental status in terms of positive, neutral or negative status of a horse in full action remains a challenge. This work aims to classify the mental status of a physically active horse as positive, neutral or negative in real-time by using wearable technology. Ten horses participated in an experiment where each horse was lunged at walking speed for 5 minutes and then at trotting speed for 10 minutes. The total of 15 minutes was then immediately repeated a second time. Halfway through each of the two trotting segments, a mental response was induced to the horse, either positive or negative, and in a random order to know when a certain mental status could be expected and reduce order effects. To induce a positive mental response, the experimenter shaked with a bowl of food (concentrate and carrots) at the edge of the lunging circle. The negative mental response was induced by means of opening and closing an umbrella abruptly, to evoke fear. Transitions to walk or trot were gently asked by voice to avoid additional stress in the horse. Heart rate and activity were continuously recorded during the experiment and each experiment was completely captured on film. Twenty-seven experiments in total were done and mental events were automatically identified using the Cumulative Sum (CUSUM) control chart algorithm. Each event was labelled as positive, neutral or negative based upon audio-visual scoring of the behaviour of the horse on the recorded videos. The classification performance of a linear and quadratic discriminant classifier, and a neural network were investigated by stratified 10-fold cross-validation. The neural network performed best on a dataset of 123 labelled mental events in terms of accuracy, with a value of 0.7182. The results indicate that it must be possible to classify mental events as positive, neutral or negative based on measurements of heart rate and activity.

The use of selective serotonin reuptake inhibitors to try to develop a pharmacological model of positive affect in sheep

Andrew Fisher[1], Leigh Atkinson[1], Emily Houghton[2], Caroline Lee[3] and Rebecca Doyle[1]
[1]The University of Melbourne, Animal Welfare Science Centre, Faculty of Veterinary and Agricultural Sciences, Victoria 3010, Australia, [2]Cardiff University, School of Biosciences, Cardiff, CF10 3AT, United Kingdom, [3]CSIRO, Animal Behaviour & Welfare, Locked Bag 1, Armidale NSW 2350, Australia; adfisher@unimelb.edu.au

There is growing interest in animal welfare research in understanding and measuring positive emotional states in animals. To enhance future methodological research on improving measurement of positive affect in animals, the aim of this study was to develop a pharmacological model of positive affect in sheep. The hypothesis was that sheep treated with fluoxetine or citalopram would have a more positive affective state and thus be more likely to approach ambiguous bucket positions than control sheep. Following approval by the University Animal Ethics Committee, 32 female Merino sheep (9-18 months of age and initially naïve to the testing facility) were trained and evaluated to have successfully and individually learned a cognitive bias testing procedure in which approach to a positive feed bucket location was rewarded by feed present in the bucket, whereas a negative bucket location had no feed present and a loud buzzer sounded if approached. Animals were then randomly allocated to treatments: (1) Control saline injection subcutaneously 30 min before testing; (2) Fluoxetine 40 mg administered orally on day -2 and day -1 and saline injection 30 min before testing; (3) Citalopram 0.5 mg/kg bodyweight injected 30 min before testing; (4) Citalopram 0.2 mg/kg bodyweight injected 30 min before testing. Each sheep was exposed to three of the four treatments (i.e. 3 testing days), with a 1-week period between treatments. Cognitive bias testing measured sheep go/no-go behaviour and latency to approach the two training bucket positions (negative and positive) and three intermediate ambiguous positions. Latency data were analysed using survival analysis with a Cox's proportional hazards mixed effects model. Results showed that bucket position significantly affected latency to approach (P<0.001), with approach behaviour declining incrementally as bucket positions became close to negative. Day of testing effects revealed that sheep were more likely to approach ambiguous bucket positions on Day 1 (66%) than on Days 2 and 3 (53% and 51%, respectively; P<0.007). There were no effects of treatment on approach responses (P=0.71), although a treatment × day effect was present (P=0.013), whereby citalopram at 0.5 mg/kg caused sheep to approach all bucket locations more on Day 2 than Day 1, whereas sheep on other treatments showed a decline in approach from Day 1 to Day 2. Approach to the positive bucket was consistent, suggesting pharmacological changes in appetite were not mediating the effects recorded. In conclusion, treatment with fluoxetine and citalopram at the dosages used in this study did not appear to induce a positive judgement bias in Merino sheep. Further work on pharmacological agents and doses, or development of alternative, behaviourally-based models may need to be undertaken in order to provide validated models of positive affect in sheep.

Video footage captured in a walk-over-weigh (WoW) system can be used to assess welfare state in sheep

Emily Grant, Amy Brown, Sarah Wickham, Fiona Anderson, Anne Barnes, Patricia Fleming and David Miller
Murdoch University, School of Veterinary and Life Sciences, 90 South Street, Murdoch, 6150, Australia; e.grant@murdoch.edu.au

Extensive sheep farmers often don't have the time or resources to regularly monitor the health and wellbeing of their animals. Recent technological advancements in automated data capture have made this task easier, though to date this technology has not included behavioural monitoring. We tested whether quantitative and qualitative behavioural assessment (QBA) methods could be used to assess sheep in different welfare states as they traversed a walk-over-weigh (WoW) system. Video footage was remotely collected from thirty-six Merino sheep within four treatment groups; control (n=12), habituated (n=8), lame (n=8) and inappetant (n=8) as they traversed the WoW system. The habituated sheep were exposed to a low-stress handling regime for six consecutive days prior to filming. At the same time, the feeding behaviour of 870 electronic ID tagged sheep at an automated feeder was recorded and those animals that had an average intake more than two standard deviation units below the population mean over the six days were designated as inappetant. A lameness scoring system was also employed to identify lame individuals, and the control animals were selected ensuring that they were not lame, inappetant or habituated. The footage from these 36 sheep was presented at random to 18 observers, blind to the treatments, for QBA analysis. Using a free choice profiling approach, observers used their own descriptive terms to score the behavioural expression of each animal along a visual analogue scale. Footage was also used to measure quantitative behavioural traits. There was a high level of agreement between the observers (P<0.001), with the generalised Procrustes analysis (GPA) consensus profile explaining 58.2% of the variation between observer scores. Two main dimensions of behavioural expression were identified from this consensus, explaining 73% and 9% of the variation between sheep scores, respectively. Treatment differences were identified on this first dimension (P<0.05); habituated and lame sheep consistently scored higher (more focused/collected/assured) compared to the control sheep (more reluctant/tense/wary). There was no difference between the inappetant sheep compared to the controls. No treatment differences were observed on GPA 2. There was agreement between the qualitative and quantitative behavioural measures. Those sheep that baulked more frequently at the entrance to the WoW system (R_s=-0.70; P<0.001) or had a higher number of circling incidences (R_s=-0.68; P<0.001) were described as more reluctant/tense/wary, and those that recorded higher walking speeds (R_s=0.65; P<0.001) or spent less time standing stationary (R_s=-0.48; P<0.01) were more focused/collected/assured. In conclusion, differences in welfare states (lame and habituated vs control) can be identified from video footage as sheep traversed the WoW system. These findings suggest that behavioural measures could be collected remotely to provide meaningful assessments of sheep welfare.

How does a stational cow brush affect cattle physiologically ad behaviourally?

Daisuke Kohari, Kenji Shimizu and Tsuyoshi Michikawa
Ibaraki University, College of Agriculture, Ami 4668-1, Inashiki-gun, Ibaraki, 300-0331, Japan;
daisuke.kohari.abw@vc.ibaraki.ac.jp

Cow brushes are used as an enrichment device to facilitate self-grooming of cattle. The degree to which they are used cattle and the brushing areas of body parts have been well reported from many studies. However, few reports of the relevant literature describe studies of brush use effects on cattle. To clarify brush effects on cattle, we investigated their heart rate variability and their behaviour when they used it. Experiments were conducted at experimental pens (3×4 m) at the Field Science Center of Ibaraki University. Five beef steers observed for this study were housed alone in the experimental pen during test days. A stationary cow brush (B2; DeLaval Inc.) was set at the side fence of the pen, adjusted to the height of their withers. Before observation, a heart rate monitor transmitter (RS300CX; Polar) was attached to each animal. Behavioural observations were conducted for 2-6 h per day for each animal. We observed them for 10.6-24.4 h during 22 days. Brush grooming durations were recorded by continuous sampling. The average heart rate and root mean square successive difference (rMSSD) between adjacent R waves of heart rate variables were calculated 30 s before, during, and 30 s after each bout of brush grooming. Furthermore, maintenance behaviour (standing or lying resting, grazing, self-grooming, brush grooming, defecation, others) and social behaviour (agonistic interaction or social grooming with cattle in neighbouring pens) of each animal were observed every 1 min using instantaneous sampling. Differences of the average heart rate and rMSSD before, during, and after brush grooming were analysed using repeated ANOVA for each animal. Also, the brush grooming duration was compared when the average heart rate or rMSSD was increased and decreased from before to after brush grooming using t-tests.. Furthermore, data of maintenance and social behaviours immediately before and after brush grooming were extracted from all behavioural data and their percentages were calculated. We observed 98 bouts constituting 1,015 s of brush grooming during 81.9 h observation. Neither the average heart rate nor rMSSD of any steer was significantly different before, during, or after brush grooming. The brush grooming duration was not different with the average heart rate transitions from before to after brush grooming. However, the brush grooming duration was significantly longer in rMSSD increased from before to after brush grooming (41.4±26.3 s) than in rMSSD decreased (24.7±20.3 s) (t=2.17, P<0.05). Before brush grooming, the most frequently observed behaviour was standing resting (61.5%) followed in rank by self-grooming (17.9%). Self-grooming was also the most frequently observed behaviour before standing resting: 31.8%. The most frequently observed behaviour after brush grooming was standing resting (57.9%), followed in rank by grazing and lying resting (13.2%). These results suggest that the brush grooming duration affects calming levels and that cattle might indicate satisfaction behaviour by resting after brushing.

Key-note:

The affective dyad: the role of maternal behaviour in positive affective states of mother and young

Cathy M. Dwyer

SRUC, Animal and Veterinary Sciences, Roslin Institute Building, Easter Bush Campus, Edinburgh, EH25 9RG, United Kingdom; cathy.dwyer@sruc.ac.uk

Maternal care provides offspring with nutrition, warmth, protection, comfort and opportunities for social learning, which serve to promote the survival of the offspring. In farmed mammalian livestock, the first two benefits of maternal care are well recognised, even in early weaning systems. However, the other attributes are less recognised and provision for these less likely to be considered in management systems. Further, any benefits of providing maternal care for the welfare of the mother are rarely considered. The onset of maternal care in mammals is associated with a complex interplay of foetal signals, steroidal hormones and the release of oxytocin at parturition. Oxytocin is also stimulated in the mother with suckling interactions and mother-offspring contact. Oxytocin has been shown to be associated with positive emotional states in humans and repeated oxytocin treatment decreases anxiety in rats. In addition, ventral contact between mother and pups has also been shown to reduce maternal anxiety, with anxiety increasing the longer mothers are away from their pups. In sheep and rats, where these processes have been well studied, parturition is also associated with alterations in dopamine signalling which may reflect rewarding properties of interacting with offspring. It seems plausible, therefore, that the expression of maternal care in mammals is associated with positive emotional states and could be considered as a component of positive welfare. For the offspring, mother-rearing shapes behavioural development, influencing social and environmental preferences, and reduces expression of fear, stress and pain responses. Mothers have also been shown to respond to pain expression in their offspring, for example increased maternal attentiveness when lambs were castrated or tail docked. The amount of maternal care expressed correlates with active pain behaviours of the lamb, although whether this contributes to decreasing pain experience of the young is not known. Thus, it is well known that maternal care contributes to the positive welfare of the offspring. However, whether these benefits can only be achieved by the biological mother, or if similar effects can be elicited by the presence of other conspecifics, or a human care giver, still needs further investigation. With the increase in interest in the importance of positive emotional states in animals, management systems that can provide prolonged periods of mother offspring contact may be important sources of positive welfare for both mother and young.

Dairy cows and heifers prefer to calve in a bedded pack barn or natural forage compared to open pasture

Erika M. Edwards[1], Katy L. Proudfoot[2], Heather M. Dann[3] and Peter D. Krawczel[1]
[1]The University of Tennessee, 2506 River Drive, Knoxville, Tennessee 37996, USA, [2]The Ohio State University, 1920 Coffey Road, Columbus, Ohio 43210, USA, [3]William H. Miner Agricultural Research Institute, 1034 Miner Farm Road, Chazy, New York 12921, USA; eedwar24@vols.utk.edu

Despite years of research, the incidence of disease after calving in dairy cattle remains stagnant. Designing a calving environment that accommodates cow preferences and natural calving behaviours may allow cows to better cope, thereby improving calving outcomes. Cows will seek a secluded area to calve when housed alone, but this behaviour has not been established in group-housed cows. The objectives were to determine dairy cows' preference for calving location when grouped and provided access to a natural environment, and to evaluate the relationship between cows' activity during the last 21 d of gestation and preference for calving location. Multiparous Holstein dairy cows (n=33) and nulliparous Holstein heifers (n=34) were enrolled together as a dynamic group with size ranging from 21 to 1, as cows were added once weekly and removed after calving. Cows were housed in a bedded pack barn with free access to 2.06 hectares of pasture beginning 21 d before expected calving date. The bedded pack barn was 13.94 m^2 and included 22 headlocks where cows were provided a TMR daily. From -21 d until after calving, cows had access to the bedded pack (section 1), open pasture subdivided into areas of approximately equal size (~0.23 hectares; sections 2 through 8), and an area of natural forage at the far end of the pasture (section 9). Man-made hides (n=6) were created to provide an alternative option for seclusion in the pasture; hides were rotated among sections 3 to 8 every week. Video data were collected to determine calving location. Dataloggers were attached to each cows' rear fetlock 21 d prior to their expected calving date to record steps (n/d). To determine preference for a calving location, data were analyzed using a chi-square test in SAS. To determine the relationship between steps and calving location, linear regression was used. Twenty-three calvings (34%) occurred on the bedded pack, and 25 (37%) calvings occurred in the natural forage area. The remaining 19 calvings (28%) occurred in one of the 7 sections of open pasture, and of these, 3 cows calved within close proximity to the hides. There was a preference for the bedded pack barn (P=0.005) and natural forage area (P=0.002) for calving over the open pasture. Within sections, heifers calved in the area of natural forage more often than cows (P=0.03), and cows tended to calve in the barn more often than heifers (P=0.06). There was no relationship between the mean steps taken daily during the 21 d prior to calving and calving location (3,512.7±44 n/d (overall mean ± SE); P=0.60; R^2=0.0045). The results provide evidence that both heifers and cows seek seclusion in a group setting when provided access to a covered barn and an area of natural forage. Cows preferred the covered barn to calve, whereas heifers preferred natural forage. This difference may be driven by previous experience; cows had greater exposure to indoor housing, whereas heifers had only been housed on pasture.

Factors affecting maternal protective behavior in Nellore cows

Franciely Costa[1,2], Tiago Valente[1], Marcia Del Campo[3] and Mateus Paranhos Da Costa[1,4]
[1]*ETCO Group, Via de Acesso Prof. Paulo Donato Castellane s/n, 14884-900, Jaboticabal, Sao Paulo, Brazil,* [2]*Postgraduate Program in Animal Science, FCAV, State University of Sao Paulo (UNESP), Department of Animal Science, Via de Acesso Prof. Paulo Donato Castellane s/n, 14884-900, Jaboticabal, Sao Paulo, Brazil,* [3]*National Program of Meat and Wool, INIA, Ruta 5, Km 386, Tacuarembo, Uruguay,* [4]*FCAV, State University of Sao Paulo (UNESP), Department of Animal Science, Via de Acesso Prof. Paulo Donato Castellane s/n, 14884-900, Jaboticabal, Sao Paulo, Brazil; mdelcampo@tb.inia.org.uy*

The aim of this study was to evaluate the effect of different factors on maternal protective behavior (MPB) of Nellore cows. A total of 3629 cows were evaluated around 24 h after calving, while their newborns were handled for navel care, ear tattoo and weighing. MPB was assessed by observing each cow reactions when its calf was caught in a corral pen. MPB score consisted of 5 levels, going from 1 (the cow paid no attention to the calf and remained indifferent to the procedure) to 5 (the cow showed aggressive behaviors toward the handler, trying to attack). The cow body condition was registered using the scale from 1 to 5. Information regarding cow age at calving was available from the farm and six classes were defined (1: one to two years old, 2: two to three years old, 3: three to four years old, 4: four to five years old, 5: five to six years old, and 6: six to eight years old at calving). A generalized linear mixed model was fitted to evaluate the effect of cow body condition, cow age at calving, calf sex and calf birth weight on MPB. The cow body condition score, its age at calving and the calf sex were considered as fixed effects. Additionally, calf birth weight was used as a covariate and the cow was included as a random effect in the model. Cows with the greatest body conditions were more protective with their calves (F=9.79, P<0.0001). Younger cows were less protective to their offspring (F=4.66, P=0.0003). Adjusted means ± SE of MPB for each cow age class were: 2.41±0.06 (1), 2.56±0.04 (2), 2.62±0.03 (3), 2.67±0.04 (4), 2.73±0.06 (5) and 2.86±0.11 (6). The calf sex had a significant effect on MPB (F=6.25, P=0.0126) being higher for mothers of females calves (2.68±0.03) compared to male calves (2.60±0.03). Results from this study also showed that cows were more protective towards heavier calves at birth (F=19.05, P<0.0001). In conclusion, older Nellore cows, with a better body condition at calving, with females and heavier offsprings showed more intensive maternal protective behavior toward their newborns.

Preference for mother does not last long after weaning in lambs

Rodolfo Ungerfeld[1], Aline Freitas-De-Melo[1], Raymond Nowak[2] and Frédéric Lévy[2]
[1]Facultad de Veterinaria, Universidad de la República, Lasplaces 1620, Montevideo 11600, Uruguay, [2]UMR INRA, CNRS, Université F. Rabelais, IFCE, 37380 Nouzilly, 37380 Nouzilly, France; rungerfeld@gmail.com

Our hypothesis was that although lambs prefer their mother after weaning, this preference does not last long. The aim of this experiment was to study the effects of various durations of mother-young separation after weaning on the maintenance of a preference for the mother. Eighty-two single Corriedale lambs stayed with their mother until weaning at 3 months of age. Preference for the mother was tested using a two-choice test between their own and another mother from the same flock, matched for age and physiological status. Lambs were tested in a 10×10×7 m triangular testing enclosure with one ewe placed in each of the two holding pens. A plastic fence line prevented lambs from approaching the ewes at less than 1 m. A proximity zone near the ewes was marked in the floor (1×4 m). The location of the mother and the other ewe in the holding pens was alternated between tests. The total time spent and the number of vocalizations in the proximity zone near each ewe was recorded for 3 min. Lambs were habituated to this enclosure before testing. All the lambs were tested 3 days before weaning, and 30 (n=25), 40 (n=15) or 60 days (n=42) after weaning. Before weaning, lambs vocalized more frequently (16.2±1.2 vs 3.4±0.5, respectively; P<0.0001) and stayed in the contact zone for longer (104.1±25.3 s vs 20.3±4.2 s respectively; P<0.0001) when near their mother than near the other ewe. Thirty days after weaning, lambs still vocalized more and stayed in the proximity zone of their mother for longer than in that of the alien ewe (vocalizations: 5.9±1.5 vs 1.5±0.6; P=0.005; time in the zone: 36.6±9.5 s vs 13.4±4.8 s; P=0.016). By contrast, when tested 40 or 60 days after weaning there were no differences in any of these variables. The decrease in the number of vocalizations emitted by the lambs in the contact zone of their mother was more pronounced in lambs tested 40 and 60 than in those tested 30 days after weaning (16.9±2.9, 15.2±1.8, and 9.1±2.4, respectively P=0.04). There was no difference in the number of vocalizations emitted in the proximity zone of the other ewe between the three testing periods after weaning (2.1±0.9, 2.0±1.9, and 1.7±1.0). Before weaning, when lambs were almost 3 months old, they had a clear preference for their mother than for an alien ewe. However, that preference decreased 30 days after weaning and disappeared completely 10 days later. Although it may be expected that according to their age, lambs tested 40 and 60 days after lambing can easily recognize their mother' face, they failed to prefer her toward another known ewe coming from the same flock. Considering that though it has not been studied in lambs, ewes can remember faces from other individual ewes during at least 2 years, this implies that although lambs can recognize her, they lost the motivation to maintain the preference for their mothers shortly after weaning. Overall, the preference of lambs for their mother fades rapidly after 30 days of mother-young separation suggesting that attachment to the mother does not last long after weaning.

Behavioural responses of dairy calves to separation: the effect of nutritional dependency on the dam

Julie Føske Johnsen[1], Cecilie Mejdell[1], Annabelle Beaver[2], Anne Marie De Passille[2], Jeffrey Rushen[2] and Dan Weary[2]
[1]*Norwegian Veterinary Institute, Department of health surveillance, P. O. Box 750, 0106 Oslo, Norway,* [2]*University of British Columbia, Faculty of Land and Food systems, 2357 Main Mall, Vancouver, BC V6T 1Z4, Canada; julie.johnsen@vetinst.no*

Rearing a calf with the cow has health and welfare benefits, but separation and weaning are challenges. We studied how the calf's nutritional dependency on the cow affects calf behavioral response to separation. Dependent calves (D, n=10) and semi-dependent (SD, n=10) calves could suckle from their dam at night, but SD calves also had continuous *ad libitum* access to an automated milk feeder (AMF). Independent (I, n=10) calves had *ad libitum* access to an AMF and their dams wore an udder net that prevented suckling. Separation took place after six weeks. Calf responses were recorded using live observations for 4 days of partial separation (with fence-line contact with the dam) and then for 3 more days after removal of the dam from the barn (total separation). Vocalizations were classed as high-pitched (open mouth) or low-pitched (closed mouth). From d 1 of partial separation, all calves had *ad libitum* access to the AMF. We hypothesized that the calves accustomed to use the feeder before separation (i.e. SD and I calves) would continue to use it, and that calves would show fewer responses to separation if they were well established on the AMF. Differences were analyzed using non-parametric statistics and results presented as, median (25th percentile, 75th percentile). During partial and total separation, I calves produced fewer high-pitched vocalizations/d than D and SD calves combined (0.00 (0.00,0.00) vs 7.2 (0.75,23.51), Mann-Whitney U=12.2, z=-3.21, P=0.001 and 0.00 (0.00,0.00) vs0.00 (0.00,1.00), U=40.5, z=-2.25, P=0.024), and also tended to produce fewer low-pitched vocalizations during partial separation (0.00 (0.00,0.05) vs 1.17 (0.00,2.54), U=29.5, z=-1.90, P=0.057). During the separation phases, 23 calves (four, nine and ten calves of D, SD and I treatments, respectively), consumed at least 1.5 l/d from the AMF; these calves produced fewer high-pitched vocalizations/d than calves not using the AMF (0.00 (0.00,7.83) vs 8.33 (5.13,24.83), U=16.0, z=-2.219, P=0.027) and tended to produce less low pitched vocalizations (0.00 (0.00,1.67) vs 0.8 (0.00,2.91), U=21.5, z=-1.731, P=0.083). During partial separation, calves' high pitched vocalizations were strongly negatively correlated to daily milk intake, (Spearman's rho=-0.770, P<0.001) and digestible energy intake (r=-0.544, P<0.011). During total separation, calves using the AMF spent more time playing (s/d) compared to calves not using it implicating that they were separated and weaned at the same time (3.67 (0.00,29.00) vs 0.00 (0.00,0.17), U=28,0, z=-1.990, P=0.047). The results indicate that nutritional independency from the dam can reduce the behavioural responses to separation from the dam. The results also show that disentangling separation from the dam and weaning from milk can improve calf welfare at separation.

Does farrowing duration affect maternal behavior of outdoor kept sows?

Cecilie Kobek Thorsen[1], Sarah-Lina Aagaard Schild[1], Lena Rangstrup-Christensen[1], Trine Bilde[2] and Lene Juul Pedersen[1]
[1]Aarhus University, Animal Science, Blichers alle 20, 8830 Tjele, Denmark, [2]Aarhus University, Bioscience, Ny Munkegade 116, 8000 Aarhus C, Denmark; cecilie.thorsen@anis.au.dk

Piglet mortality is high the first three days after farrowing in Danish outdoor production. The use of high prolific sows is associated with long farrowing duration due to the greater number of piglets born. Prolonged farrowing may exhaust the sow and cause pain, which may affect maternal behavior and cause increased piglet mortality. The behavior of thirty-eight outdoor kept sows were video recorded from inside their individual hut during farrowing and the following three days. Farrowing duration, nursing pattern, sow posture changes, carefulness when lying down and latency to leave the hut after birth of the last piglet were recorded. The effect of farrowing duration on sow behavior was analyzed as a continuous variable using Proc Mixed in SAS. A mean farrowing duration of 7.5 h (range 2-29 h) to last live born piglet was recorded with a mean of 17 total born piglets (range 11-25). The only effect found of farrowing duration was that the latency to leave the hut after birth of the last piglet increased by one hour for every hour increase in farrowing duration ($F_{1,110}=7.37$, $P=0.008$, mean=15.7±7 h). Farrowing duration had no effect on nursing frequency, duration sows presented the udder, number of posture changes nor the risk of 'flopping' when lying down. The results, however, did show normal patterns of increased activity as the days passed. Thus, prolonged farrowing did not seem to cause exhaustion to an extent where maternal behaviour was affected.

Does dietary tryptophan around farrowing affect sow behavior and piglet mortality?

Jeremy N. Marchant-Forde[1], Brian T. Richert[2], Jacob A. Richert[2] and Donald C. Lay, Jr.[1]
[1]USDA-ARS, LBRU, West Lafayette, IN 47907, USA, [2]Purdue University, Department of Animal Sciences, West Lafayette, IN 47907, USA; jeremy.marchant-forde@ars.usda.gov

Piglet mortality remains a serious welfare and economic problem. Much of the early mortality is due to crushing by the sow. Tryptophan has been shown to reduce aggression and have a calming effect on behavior, which may reduce the number and type of posture changes, thereby altering risk of crushing. The aims of this experiment were to determine if feeding dietary tryptophan around farrowing would affect quantitative measures of posture-changing behavior and piglet mortality. Twenty-four multiparous sows (parity 2, 3 and 4) were moved to the farrowing house on d 110 of pregnancy and randomly assigned to one of two treatments: (1) standard lactation diet (CTL); or (2) experimental lactation diet with same energy content but containing four times the amount of tryptophan contained in the standard diet (TRYP). Treatments were applied from entry to the farrowing house until 3 d post-farrowing, with all sows receiving standard lactation diet thereafter. Sow behaviour was video-recorded continuously from entry until 7 d post-farrowing and all occurrences sampling was used to determine number and type of posture changes from 3 d before until 2 d after birth of first piglet. Posture changes recorded were all transitions between standing, sitting, kneeling, lying sternally and lying laterally. Production data recorded included sow weight on entry to farrowing crate, sow weight at weaning, sow feed intake, number of piglets born alive, dead and mummified, birth weight, 24 h weight, mortality and cause (physical inspection in conjunction with video data), number weaned, and weaning weight. Data were analyzed using the MIXED procedure accounting for repeated measures when appropriate. Dietary treatment had no effect on total litter size or number of piglets born alive (P=0.30), but TRYP sows tended to give birth to fewer dead piglets than CTL sows (1.0±0.4 vs 2.5±0.7, P=0.07). Total feed intake over the immediate pre- and post-farrowing period and sow lactation weight loss was not different between treatments, but on the day of farrowing, TRYP sows tended to eat more than CTL sows (3.0±0.4 vs 1.8±0.4 kg, P=0.09). Piglet birthweight, weaning weight and growth rates did not differ and total piglet mortality (born dead + liveborn mortality) was similar between treatments (TRYP 17.0±2.8 vs CTL 24.3±5.4%, P=0.29). The number and type of posture changes varied over time, but the only difference between treatments during the critical immediate 2 d post-farrowing period, was that CTL sows transitioned less between sitting and lying than TRYP sows (TRYP 34.9±5.6 vs CTL 21.2±2.0, P=0.03). Overall, feeding a high tryptophan diet around the time of farrowing had little effect on sow posture-changing behavior and no effect on liveborn piglet mortality. There may be a beneficial effect on stillbirth incidence and farrowing day feed intake, which could affect early lactation milk production, but this requires further investigation.

Relationships between sow conformation, accelerometer data and crushing events in commercial piglet production

Stephanie M. Matheson[1], Rob Thompson[2], Thomas Ploetz[2], Ilias Kyriazakis[1], Grant A. Walling[3] and Sandra A. Edwards[1]

[1]Newcastle University, Agriculture, Food and Rural Development, Newcastle upon Tyne, NE1 7RU, United Kingdom, [2]Newcastle University, Computing Science, Newcastle upon Tyne, NE1 7RU, United Kingdom, [3]JSR Genetics Limited, Southburn, East Yorkshire, YO25 9ED, United Kingdom; stephanie.matheson@newcastle.ac.uk

Pig selection for meat production maximises growth rate and length of back with a resulting change of body shape and difficulty in control of posture change in sows. As a consequence, many piglets are crushed, either as the sow lies down or rolls from side to side. This study investigated how characteristics of posture change movements in the sow (in conjunction with leg conformation) affect the likelihood of piglet crushing. Piglet births (n=11,752) were recorded for 21 weeks in a population of approx. 750 Landrace sows crossed with either White Duroc or Large White sires. All sows had leg conformation data collected (using previously published criterion), with a subset having rump-mounted accelerometer data (n=315). Data extracted from the accelerometers included the mean rate of change in movement around the X-axis (ROLL-CHANGE) and the mean rate of change of movement around the Y-axis (PITCH-CHANGE). The farrowing floors were either concrete/plastic, concrete/metal, fully metal or fully plastic. Piglet data gathered at processing (18-24 hours after birth) were piglet weight, sex, IUGR-status determined by head morphology (normal, light-IUGR and severe-IUGR) and reason for death. All piglets remained in their birth litters until processing, but were fostered thereafter. Females were individually identified but males were unidentifiable after processing. There were 349 piglets (both males and females) crushed between birth and processing (CPROC); 146 females were crushed between processing and weaning (males were unidentifiable after processing) giving a total of 495 crushed piglets (CTOT). Non-significant effects (GLMM at piglet level with sow-week as a random factor) on both CTOT and CPROC included sow parity, piglet sire breed, gender and the direct effects of sow conformation. Significant effects on CPROC were an interaction between piglet weight and IUGR-status (P=0.004), the type of flooring in the farrowing house (P=0.038), ROLL-CHANGE (P=0.043), PITCH-CHANGE (P=0.008), the interactions between PITCH-CHANGE and the hind pastern angle (P=0.039) and PITCH-CHANGE and placement of the hind feet (P=0.035). There was also an interaction between ROLL-CHANGE and the shape of the hind leg (P=0.077). For effects on CTOT, the piglet weight/IUGR-status interaction retained significance (P=0.002), as did PITCH-CHANGE (P=0.002) and the interactions between PITCH-CHANGE and the hind pastern angle (P=0.032) and PITCH-CHANGE and hind feet placement (P=0.018). The heritabilities for the accelerometer traits were: PITCH-CHANGE, h^2=0.32±0.171 and ROLL-CHANGE, h^2=0.03±0.110. In conclusion, accelerometer-derived measurements of sow movement have an effect on early piglet crushing. Sow limb conformation influences crushing events through interactions with sow movements. The interactive effects of sow conformation and accelerometer-derived data merit more detailed investigation. This research was funded by the EU FP7 Prohealth project (no. 613574).

Effects of litter size on maternal investment and neonatal competition in pigs with different genetic selection pressures

Marko Ocepek, Ruth C. Newberry and Inger Lise Andersen
Norwegian University of Life Sciences, Department of Animal and Aquacultural Sciences, P.O. Box 5003, 1432, Norway; marko.ocepek@nmbu.no

Artificial selection of the domestic pig (*Sus scrofa domesticus*) offers a useful model for investigating changes in behaviour associated with reproductive trade-offs between litter size and fitness of offspring. The aim of this study was to evaluate effects of litter size on teat stimulation, sibling competition and pre-weaning survival and growth in three populations of domestic pigs subjected to different selection pressures (a maternal line emphasizing high weaned pig production (Norsvin Landrace), a paternal line emphasizing meat traits (Norsvin Duroc), and a crossbred line (Norsvin Landrace and Yorkshire)). We predicted that, with increasing litter size, piglets would spend more time in teat massage, be less likely to gain access to a teat during milk letdown and, if surviving to weaning, have lower, more variable body weights. We also predicted that maternal line sows would wean more piglets of higher weight, despite larger litter sizes, than paternal line sows. Sows (maternal line, n=12, paternal line, n=12, crossbred line, n=14) were loose-housed with their litters in individual farrowing pens. We collected data on piglet behaviour during nursings at 1 day of age, when sibling competition was expected to be most intense. Piglets were weaned at 35 days of age, when they were weighed and cumulative mortality was calculated. Effects of 1-day litter size, genetic line and their interaction on pre- (mean ± SE, 153±9 s) and post- (552±39 s) udder massage duration, piglets missing milk letdown (10.7±1.4%), nursing interval (54±1.9 min), nursing success (92.5±1.6%), piglet mortality to weaning (10.9±0.5%), mortality due to starvation (3.2±1.3%) and crushing (4.5±1.3%), piglet weight at weaning (11.1±0.3 kg) and variation in weight at weaning (14.9±1.1%) were analysed using a general, or generalised, linear model (depending on distribution of residuals). With increasing litter size, piglets spent more time in pre- (P=0.050) and post- (P<0.001) letdown massage and had shorter nursing intervals (P=0.018). Increasing litter size resulted in larger litters at weaning (P=0.002), but at a cost of more frequent termination of nursings prior to letdown (P<0.001), more piglets without a functional teat at letdown (P<0.001), an increased risk of mortality due to starvation (P<0.001) and crushing (P=0.002), and lower (P=0.039), more variable (P=0.002) body weights at weaning. Despite more post-letdown massage in the maternal line (litter size × breed: P<0.001), nursing intervals were longer (P<0.001) and mortality due to crushing was higher (P<0.001) with increasing litter size in this line, without increments in number, weight or weight uniformity of weaned piglets with increasing litter size (litter size × breed: P>0.1). Our results suggest that continued selection pressure to increase sow reproductive capacity is socially unsustainable due to the increasing animal welfare costs resulting from increased piglet competition.

The piglet weigh test – a novel method to assess sows' maternal reactivity

Anna Valros[1], Margit Bak Jensen[2] and Lene Juul Pedersen[2]
[1]*University of Helsinki, Faculty of Veterinary Medicine, Department of Production Animal Medicine, P.O. Box 57, 00014 University of Helsinki, Finland,* [2]*Aarhus University, Department of Animal Science, Blichers Alle 20, 8830 Tjele, Denmark; anna.valros@helsinki.fi*

A so called piglet scream test has been developed to test the maternal reactivity of sows to, among others, assess effects of housing systems under experimental conditions. However, this test is laborious and difficult to perform in a large scale. We propose a simple test of sows' maternal reactivity, 'the piglet weigh test'. We related the sows' behavioural reaction during this test to heart rate (HR), and compared the maternal reactivity of sows in two farrowing environments. The study included twelve 2-3 parity sows (Landrace × Yorkshire). Ten sows were housed in farrowing crates (4.8 m²) and 2 in farrowing pens (6.6 m²). Before weighing of piglets when these were 4-6 d old, sows were fitted with a HR belt and HR (beats/min) was measured during weighing of the 5 first piglets (Piglet 1 to 5). Piglets were handled gently, but for long enough to make them vocalize for at least 5 s. Sow behaviour was observed via video and maternal reactivity scored as either 0: no reaction visible; 1: small reaction (turns head once to look at the piglet handler); 2: clear reaction (walking, headshaking, manipulating cage, getting up, or turning head several times). Statistical tests were performed with IBM SPSS 21. Changes in HR during the weighing were tested with paired t-test by comparing HR level during weighing of each of the Piglets 1 to 5 to the HR level during a resting period (Rest). Test of differences in average HR during weighing Piglet 1 to 5 between sows with different scores of maternal reactivity was performed with ANOVA, followed by pairwise comparisons, using the Bonferroni correction. Effect of housing on average HR during the piglet weighing was tested with a t-test. The maternal reactivity was scored 0 in 5 crated sows, 1 in 3 crated sows and 2 in 2 crated sows. The two penned sows scored 1 and 2, respectively. HR increased significantly during piglet weighing: sow HR was higher than at Rest (mean 85 (SD 11)), when weighing Piglet 2-4: 2: 98 (15), 3: 98 (16), and 4: 98 (15), $P<0.05$ for all. HR when weighing Piglet 1 (92 (12)) and 5 (94 (13)) tended to differ from Rest HR ($P<0.1$ for both). Average HR increased with increasing maternal reactivity (overall $P=0.02$, 89 (4.6), 92 (12), and 115 (13) for score 0, 1 and 2, respectively). A significant difference in HR was found between sows scoring 0 and 2 ($P=0.02$), and sows scoring 0 and 1 tended to differ ($P=0.05$). Average HR tended ($P=0.06$) to be higher in penned sows (114 (21)) than in the crated sows (93 (11)). Scoring the maternal reactivity to piglet handling in this simple way correlated with a physiological reaction of the sows, indicating that even subtle behaviour changes indicate an emotional reaction in the sow. Despite the small sample size, the effect of housing was in the expected direction. The piglet weigh test has potential value to evaluate sow maternal reactivity in applied situations, but needs further validation.

Background noise impacts older sow behavior in farrowing barns, a preliminary study

Nichole M. Chapel[1] and Donald C. Lay, Jr.[2]
[1]Purdue University, 125 S. Russell Street, West Lafayette IN 47906, USA, [2]USDA-ARS Livestock Behavior Research Unit, 125 S. Russell Street, West Lafayette IN 47906, USA; chapeln@purdue.edu

Piglet crushing may be attributed to a lack of sow awareness to piglet presence. Additionally, farrowing barns have constant noise present as ventilation systems are maintained 24 h and multiple sow-piglet combinations are attempting communication. Between these two interferences, the possibility exists that sow-piglet communication is being impaired. The aim of this study was to investigate if sows farrowed in quieter auditory environments are more responsive and can communicate more effectively with their piglets. Sows farrowed in 1 of 3 environments: a crate in a farrowing room with 8 other sows and constant ventilation noise (CON; n=6), a farrowing room void of any other sows with constant ventilation noise (ISO; n=6), or farrowed in a room void of sows, ventilation noise and overall background noise (QUIET; n=6). Sows were tested 24 and 48 h post-partum with a piglet removal event (Steal) and the return of the piglets to the sow (Return). Five-min Steal behavior observations started when a novel human removed a sow's entire litter. Behaviors observed during Steal included: total time looking toward the intruder (looking) and time searching for piglets defined as nosing or rooting. Ten-min Return observations included these behaviors and the total time until a nursing event following the return of the piglets. Parity of sows were recorded and grouped as either young (2nd and 3rd parity) or old (4+ parity). Data were analyzed using mixed model ANOVA with repeated measures. During Steals, ISO sows spent more time looking than CON sows (20.53±4.69%; 5.50±3.57%; respectively, P=0.02). QUIET sows did not differ from ISO or CON (QUIET=15.00±3.15%, P>0.05). Older ISO sows spent the highest proportion of time looking compared to young ISO sows (31.54±8.25%; 9.51±4.46%, respectively, P=0.05). Returns produced similar results with percent of time looking (P<0.001) and number of looks (P=0.004) being higher in older ISO sows than younger ISO sows (4.35±0.90%; 0.32±0.50%; 4.00±0.86; 0.29±0.47, respectively). Overall percent of time looking was lowest for CON sows during returns (Con=0.68±0.39%; ISO=2.34±0.52%; QUIET=2.42±0.35%; P=0.01). There was a tendency for older sows to nose their piglets more upon return (Old: 12.33±1.95, Young: 7.74±1.36, P=0.07). Finally, there was a tendency for older QUIET sows to begin nursing their piglets upon return earliest compared to CON or ISO sows (67.91±68.89 s; 332.00±94.54 s; 360.56±133.97 s; respectively, P=0.06). Overall, older sows showed the greatest response to the removal and return of piglets when farrowed in an environment without other sows. These results suggest that older sows may benefit from farrowing in an environment away from other sows. Therefore, it is suggested that more attention be given to sow-piglet communication during the farrowing period, particularly with concern given to background noises.

The relationship between maternal behavioural traits and piglet survival, can we breed for improved maternal ability in sows?

Inger Lise Andersen and Marko Ocepek
University of life sciences, Animal and Aquacultural Sciences, P.O. Box 5003, 1432 Ås, Norway; inger-lise.andersen@nmbu.no

The primary aim of our work was to find maternal behaviours important for piglet survival, to develop qualitative scores from continuous measures of those traits, and to study the relationship between maternal behavioural scores, piglet mortality and the number of weaned piglets in sows of three different sow breeds (Norsvin Landrace (n=12), Norsvin Duroc (n=12) and crossbred Norsvin Landrace × Yorkshire (n=14). The following qualitative maternal behaviours were scored as follows: nest building activities (scale from 1 to 3, increased frequency of nest building behaviour), sow communication score (COS: scale 1 to 4, denoting an increased amount of sniffing, grunting and nudging), and sow carefulness (CRS: scale from 1 to 4, increased degree of carefulness while moving around). There was a moderate positive correlation between the continuous measured NBA and the qualitative score for nest building NBS (r=0.469; polyserial correlation coefficient) as well as between the qualitative score for communication, COS, and the continuous, video-based measure of the amount of communication while standing, CS, (r=0.439; polyserial correlation coefficient), and the qualitative and quantitative scores similarly affected piglet survival. Since COS and CRS were highly correlated (r=0.883; polychoric correlation coefficient), we tested the effect of those behavioural scores separately on production parameters (proportion of dead piglets and number of weaned piglets) using two models (model 1: NBS, COS; model 2: NBS, CRS), and compared their relative predictive accuracies using Akaike information criteria (AIC) and AIC weights. Model 1 had better predictive accuracy. In models 1, piglet mortality decreased with higher NBS (Genmod procedure; P=0.004) due to less crushing (Genmod procedure; P<0.001) and more weaned piglets (GLM procedure; P=0.043). Increases in both COS and CRS were associated with lower overall piglet mortality (Genmod procedure; P<0.001), fewer crushing incidences (Genmod procedure; P<0.001) and more weaned piglets (GLM procedure; P=0.004). Additionally, a higher COS was associated with a lower proportion of starved piglets (Genmod procedure; P=0.002). Our results demonstrated that our three defined maternal behaviour scores had a significant impact on piglet survival, and therefore we tested the same scores in 900 LY-sows from 45 Norwegian loose-housed sow herds. There was a relatively low correlation between NBA, COS and CRS in the field data. The higher the CRS, the lower was the mortality of liveborn piglets (ProcMixed procedure; P<0.01) and the more piglets were weaned (ProcMixed procedure; P<0.009). COM also had similar effects on piglet mortality (ProcMixed procedure; P<0.001) and the number of weaned (ProcMixed procedure; P<0.001) piglets. There was no significant effects of NBS on piglet mortality, but an increaed NBA resulted in more piglets weaned (ProcMixed procedure; P=0.009). Due to these results, CRS ad COM as measures of maternal care ability in sows have the greatest potential to be tested in nucleus herds for calculation of genetic variation and heritability, and should be taken into account in a breeding program for more robust sows.

Pre-farrow dietary inclusion of magnesium to increase ease parturition and increase piglet survival

Kate Plush[1,2,3] and William Van Wettere[2]
[1]SunPork Solutions, PO Box 92, 5400 Wasleys, Australia, [2]The University of Adelaide, Animal and Veterinary Science, Roseworthy Campus, 5371 Roseworthy, Australia, [3]South Australian Research and Development Institute, Livestock and Farming Systems, Roseworthy Campus, 5371 Roseworthy, Australia; kate.plush@sunporkfarms.com.au

Sow discomfort around parturition can heighten stress, increasing risky behaviours for piglets like extended farrowing durations, frequent posture changes and in severe cases savaging. Magnesium supplementation in growing pigs reduces stress and aggression. The aim of this experiment was to determine if two sources of dietary magnesium improved farrowing ease, thereby increasing piglet growth and survival. Young sows (parity 0-3) were randomly allocated at farrowing crate entry (d110 gestation) to one of the following treatments; CON (n=13) fed 2.4 kg lactation sow mash daily, $MgSO_4$ (n=10) given an additional 21 g magnesium sulphate, and SUPP (n=14) given an additional 12 g magnesium rich marine extract. After parturition, all sows received lactation sow mash *ad libitum* Sows were monitored from birth of first piglet until placental expulsion for; number of posture changes, farrowing duration, number piglets born dead and alive. When each piglet was born the birth interval, degree of meconium staining, vitality score and birth weight was collected. At 24 h of age, piglet weight was collected again to estimate colostrum ingestion, and blood from an ear prick analysed for glucose concentration. Piglets were weighed at d21, and all litter mortality prior to fostering (12-24 h within treatment), and to weaning was noted. Most sow measures were analysed using a general linear model but litter mortalities were analysed using a generalised linear model with Poisson distribution in SPSS. Piglet measures were analysed by linear mixed model with piglet as the repeated term. There were no treatment effects on sow posture changes during parturition, farrowing duration, piglets born alive or born dead. Piglets from the SUPP treatment experienced the longest birth interval (17.0±1.6 min) compared to the other treatments (CON 11.8±1.9 min, MGSO4 13.4±1.8 min; P<0.05). No treatment effect on the degree of meconium staining was detected, but piglets from MGSO4 sows recorded a higher vitality score after birth (2.2±0.1) than CON (1.9±0.1) and SUPP (1.9±0.1; P<0.001). Weight gain to 24 h was greater for SUPP (121±11 g) than CON piglets (89±13 g) with MGSO4 (108±13 g) intermediate (P<0.05). Blood glucose concentration at 24 h was highest in MGSO4 (8.0±0.2 nmol/l) and SUPP (7.5±0.2 nmol/l) piglets and lowest in CON (6.9±0.2 nmol/l; P<0.001). More piglets died prior to fostering on CON sows (0.8±0.3, MGSO4 0.2±0.1, SUPP 0.2±0.1; P<0.05), but there we no treatment effect on mortality after this time. Number of piglets weaned was unaffected by treatment but piglets from MGSO4 sows tended to weigh less at d21 (6.3±0.2 kg) than CON (6.6±0.1 kg) and SUPP (6.7±0.1 kg; P=0.09). These data provide evidence that maternal magnesium supplementation reduces piglet death in the peri-natal period, and this appears to be attributed to improved energy acquisition by the piglets.

Pre-weaning environmental enrichment increases play behaviour in piglets (*Sus scrofa*) at a large scale commercial pig farm

Chung-Hsuan Yang[1], Heng-Lun Ko[2], Laura Salazar[1], Irene Camerlink[3] and Pol Llonch[2]
[1]University of Edinburgh, Royal (Dick) School of Veterinary Studies, Easter Bush, Midlothian, EH25 9RG, Edinburgh, United Kingdom, [2]Universitat Autònoma de Barcelona, School of Veterinary Science, Bellaterra, 08193, Cerdanyola del Vallès, Spain, [3]Animal and Veterinary Sciences Research Group, Scotland's Rural College, West Mains Rd., EH9 3JG, Edinburgh, United Kingdom; forseika@gmail.com

Environmental enrichment is a legal requirement for both sows and fattening pigs in Europe although not for neonatal piglets. The effect of different enrichment material in the neonatal environment has been explored extensively, but the suitability of different enrichment materials deserves more research attention. Here we investigate the effect of hanging objects vs substrate as enrichment material in the neonatal environment at a large scale commercial pig farm in Spain. Conventional farrowing pens either had no enrichment (CON, n=16 litters), or were attributed with a white plastic box (60×40×10 cm) with crushed wood substrate (SUB, n=16), or with six hanging objects (two rubber balls with a string of ropes, two wired thick ropes and two rubber cones; TOY, n=16) from 1 d after birth until weaning (d 26). Litter size was homogenized to 13-14 piglets within 24 h after birth. Play behaviour, aggression and growth were studied in up to 596 piglets from birth until 36 days of age. Locomotor, social and object play were recorded weekly by scan sampling. Aggression was assessed by the number of skin lesions on focal piglets, and individual weights were recorded weekly. Data were analysed using mixed models accounting for litter effects. The SUB and TOY groups showed 1.8 times more object play than CON piglets (P<0.001) during lactation. However, the SUB and TOY group did not differ in their amount of object play (P=0.51). Object play linearly increased with piglet age throughout lactation (P<0.001). TOY showed more social play than SUB (4.8 vs 4.3% of observations; P=0.04) but did not differ from CON (4.6% of observations; P=0.64). Social play tended to increase with age (P=0.06), with 4.8% of observations spent on social play at 12 days of age. Locomotor play did not differ between treatment groups (P=0.86) and was unaffected by age (P=0.21). Piglets had on average 0.2±0.1 lesions pre-weaning; 2.5±0.3 lesions at D1 post-weaning; and 0.7±0.1 lesions at D2 post-weaning (no differences between treatments, P>0.10). Body weight and growth rate (average 0.15 kg/d) did not differ between treatment groups throughout the trial (P>0.10). Environmental enrichment (TOY and SUB) increased object play in neonatal piglets compared to a barren environment and among the two enrichment materials, objects better promoted social play compared to substrate. However, improvements on play behaviour were not translated into greater growth performance in this study. The current study showed that providing objects as enrichment can be an effective way to enhance piglet welfare.

Strategic supply of straw for nest-building to loose housed farrowing sows reduced number of stillborn piglets

Kari Baekgaard Eriksson[1], Vivi Aarestrup Moustsen[2] and Christian Fink Hansen[1,2]
[1]University of Copenhagen, Department of Large Animal Sciences, Groennegaardsvej 2, 1870 Frederiksberg C, Denmark, [2]SEGES Danish Pig Research Centre, Axeltorv 3, 1609 Copenhagen V, Denmark; vam@seges.dk

The effects of providing large amounts of straw (strategic use of straw) on the floor 1-3 days before expected farrowing in loose farrowing pens on early piglet mortality and the sows' movement behaviour around farrowing were investigated. A total of 559 Danish Landrace × Yorkshire sows and their litters were used in the study. All sows had *ad libitum* access to straw provided in a straw rack. Sows were randomly assigned to one of three treatment groups: (1) Control (n=187) (CON); (2) an additional 5 kg short cut straw provided on the floor (n=186) (5KG); or (3) an additional 10 kg of straw provided on the floor (n=186) (10KG). From 67 sows, postures (lying ventrally, lying laterally, sitting and standing/walking) from 2 h before birth of first piglet (BFP) until 8 h after BFP were recorded and analysed. As planned, the average total amount of allocated straw differed significantly between treatment groups (CON=2.2 kg, 5KG=8.2 kg, 10KG=13.1 kg, P<0.001). In addition, the average amount of allocated straw in the *ad libitum* straw rack was higher in the CON-group compared to the two other treatment groups (2.2 vs 1.9 vs 2.0 kg, P<0.001). Litter size was not influenced by treatment and averaged 18.5 total born. Straw significantly reduced number of stillborn per litter. Compared to CON-sows the number of stillborn per litter decreased by 0.3 (P=0.045) and 0.5 (P<0.001) piglets in 5KG- and 10KG-sows respectively. The number of liveborn deaths increased in 5KG-sows by 0.2 (P=0.009) compared to CON-sows, but no difference was observed between 5KG- and 10KG-sows (P=0.275) or CON- and 10KG-sows (P=0.124). Strategic use of straw had little effect on sows' movement behaviour. During the last 2 h prior to BFP the CON- and the 10KG-sows stood up/walked for 16 and 11 minutes respectively. For the first 8 hours after BFP the sows lay laterally in approximately 90% of the time regardless of treatment. Since an effect of straw was seen in the stillbirth rate, a straw effect on sow behaviour might be found in the nest-building phase, which needs further investigation. Finally, the large amount of straw caused problems with the manure system as it accumulated in the vacuum-liquid based slurry channels. In conclusion, strategic use of straw significantly reduced the number of stillborn per litter, however, development of housing systems and especially flooring and manure systems able to handle large amounts of straw is crucial for the method of strategic use of straw to be feasible for commercial productions.

Agression, vocalization and underweight in piglets born from gilts with lameness

Marisol Parada[1,2], Thiago Bernardino[1], Patricia Tatemoto[1] and Adroaldo Zanella[1]
[1]University of São Paulo, Faculty of Veterinary Medicine and Animal Science, Department of Preventive Veterinary Medicine and Animal Health, Av. Duque de Caxias Norte, 225 CEP 13635-900, Pirassununga, SP, Brazil, [2]National University of Colombia, Faculty of Veterinary Medicine and Animal Science, Street 45 N ° 26-85, 481 Building, Bogotá D.C, 11001, Colombia; adroaldo.zanella@usp.br

Problems in the locomotor system are common in pregnant gilts and its occurrence can cause pain and stress which compromise their welfare. Lameness is a multifactorial condition, caused by nutritional factors, floor abrasiveness, social instability, among other factors. Glucocorticoids secreted in pregnant gilts under stressful conditions, can cross the placental barrier, affecting brain structures that modulate affective states and social behaviour of the offspring. Gait scores of 22 pregnant, group-housed gilts were assessed six times with intervals of two weeks between evaluations. Scores from 0 to 5 were assigned to each gilt according to a previously validated scoring system. A score of 0 represented gilts without lameness and a score of 5 represented gilts with the highest lameness score. At 107 days of pregnancy gilts were transferred to individual farrowing pens. Lame gilts were treated according to a standard protocol (Flunixin Meglumine for 4 days and antibiotic). At 28 days of age, piglets were weaned, weighed and distributed in four uniform body size groups. Lesions resulting from fights were counted in each group on days 28, 29 and 30. Two days after weaning, vocalization and activity levels were evaluated in each group using an open field and novel object tests. For the statistical analysis, piglets were grouped according to the degree of lameness of their mother. Group A corresponded to piglets (P) of mothers (G) with lameness 0 and 1 (n=8G; n=58P), group B to piglets of mothers with lameness 2 and 3 (n=9G; n=60P) and group C to piglets of mothers with lameness 4 and 5 (n=5G; n=38P). Data analysis was performed depending on the normality of the variables through ANOVA or Kruskal-Wallis tests at a significance level of 5%. Pairwise comparisons were performed using Tukey and Kramer (Nemenyi) test with Tukey-Dist approximation for independent samples. All analyzes were performed in the programming language R. Group A piglets had fewer lesions at days 28 ($P<0.05$) and 29 ($P<0.05$) than group B piglets. Although the litter size was not significantly different between groups ($P=0.223$), group B piglets were the heaviest ($P<0.05$) and group C piglets were the lightest ($P<0.05$). Additionally, group A piglets vocalized more than group B piglets ($P<0.05$) when they were subjected to the novel object test. We conclude that lameness in pregnant gilts have negative effects on the offspring, affecting weight gain, increasing aggressiveness and altering vocalization.

Performance of non-maternal as well as maternal nest-building behaviour improves reproduction in female mink

Toke Munk Schou and Jens Malmkvist
Aarhus University, Department of Animal Science, Behaviour and stressbiology, Blichers alle 20, 8830, Denmark; tokeschou@anis.au.com

Hindrance from performing motivated behaviour can induce stress. Studies have documented that mink dams are motivated for nest-building early during the gestation period. Access to suitable nesting material in the cages reduced the cortisol concentration in pregnant dams. Stress during gestation is known to negatively affect reproduction; however, the timing for onset of the nest-building motivation is currently unstudied in mink (*Neovison vison*). Thus, we aimed to investigate (1) the time for onset of nest-building in dams used for breeding, (2) whether nest-building material (straw) reduced the occurrence of abnormal behaviour, and (3) whether prolonged access to straw (up to 3.5 months prior to delivery) influenced the reproductive success. We included 126 first-year brown mink dams from January 15 into two equally sized treatment groups under identical housing and mating conditions; however, group Straw15Jan was provided with free access to straw in the cage from January 15 while this happened March 23 (just after mating) for group Straw23Mar. All dams had a top-layer of straw on the nest box which they could pull through a wired mesh. Dams, who were not successfully double mated or found barren, were excluded from the experiment from March 24. Day of birth (Day 0 – in average April 28) was registered and kits were counted and weighed Day 1 and 7 after birth. Both groups of dams had a very low nest score (visually scored on a 0 to 5.5 scale, 0 = no nest; 5.5 = complete nest with thick top layer) January 14; however, already the day after allocation of nesting material in the cages for group Straw15Jan, the nest score increased drastically with increasing nest score the following weeks. Thus free access to straw in the cage greatly increases the mink dams' nest-building (measured by nest-score). Due to nest-building from January 15 we were not able to determine the onset of maternal-related nest-building. Instead, the results show that mink dams are motivated for nest-building in the winter time (even before mating) and through to parturition, which is a new finding. Dams in Straw23Mar built a nest of similar nest score as the dams in Straw15Jan within one day when given access to straw – and the nest score remained similar. Early access to straw did not affect the number of individuals that performed stereotypy (58-68%; P>0.65), observed directly nine times of one minute, in 2 hours prior to feeding, and fur chewing (44-60%; P>0.37), scored visually as fur damage, on two test days (χ2-test). From the groups Straw15Jan and Straw23Mar, 42 and 41 dams gave birth, respectively. Generally, Straw15Jan performed better in all reproduction measures (though P>0.05 in ANOVA): litter size at birth (8.6±0.32 vs 8.2±0.34; P=0.45); stillborn kits (8.8±1.98 vs 13.5±2.94%, P=0.15); kits alive Day 1 (7.4±0.35 vs 6.5±0.38; P=0.082) and liveborn mortality until Day 7 (17.4±3.38 vs 24.4±4.24%; P=0.20). These effects summarized in a significant larger litter size Day 7 (6.6±0.38 vs 5.4±0.37; P=0.024). In conclusion, prolonged access to straw in the cage improves the early reproductive success in female farm mink.

Being reared by ewes that grazed in different pasture allowances does not affect the lamb response to a short-term separation

Aline Freitas-De-Melo[1], María Belén López-Pérez[1], Florencia Corrales[1], Camila Crosa[1], Dermidio Hernández[1], Rodolfo Ungerfeld[1] and Raquel Pérez-Clariget[2]
[1]Facultad de Veterinaria, Universidad de la República, Lasplaces, 1620, 11600, Uruguay,
[2]Facultad de Agronomia, Universidad de la República, Garzón, 780, 12400, Uruguay;
alinefreitasdemelo@hotmail.com

Lambs reared by ewes that grazed on low pasture allowance during gestation establish a weaker attachment with their mother and present an earlier nutritional independence. Thus, we tested the hypothesis that they also present a different emotional response to short-term separation from the dam. We compared the emotional response to the short-term separation of the lambs from their mother that grazed on high or low native pasture allowance (NPA) from 30 to 143 days of gestation. Forty-one multiparous single-lambing Corriedale ewes [body weight (BW): 48.0±1.1 kg; body condition score (BCS): 3.52±0.05] were randomly assigned to two NPA: (1) high (group HPA: n=22): ewes grazed on 14 to 20 kg of dry matter/100 kg of BW per day; and (2) low (group LPA; n=19): ewes grazed on 6 to 10 kg of dry matter/100 kg of BW per day. Each treatment had three replications separated by electrical fences. From 100 days of gestation until lambing ewes were daily supplemented with 300 g/animal of rice bran provided collectively to avoid risk of toxaemia. From 143 days of gestation until the test all ewes grazed in the same native pastures with unlimited availability. Ewes' BCS and BW were recorded at lambing. Lambs were weighed at birth and the day after the test. Three months after lambing, lambs were subjected to a short-term separation test in a pen (5×3 m) divided in two similar areas by a plastic fence. A reference line was drawn 80 cm parallel to the fence. All lambs were separated from their mother and housed together 30-60 min before the test. Each lamb was isolated in its area for 5 min, and then its' mother was introduced in the other side of the fence. During 5 min the number of times that the lamb crossed the line, and thus, was <80 cm from the fence, the time each lamb was nearer 80 cm from the fence, the number of times it vocalized and the number of attempts to join with their mother by hitting against the fence was recorded. At lambing, HPA ewes had greater BCS and BW than LPA ewes (BCS: 3.27±0.06 vs 2.81±0.17; BW: 52.0±0.53 kg vs 47.4±0.6 kg; P<0.0002 and 0.0001, respectively). Lambs reared by HPA ewes were also heavier than those born from LPA ewes (4.7±0.1 kg vs 4.1±0.1 kg; respectively, P=0.004). One day after the test, lambs reared by HPA ewes tended to be heavier than those reared by LPA ewes (19.0±0.7 kg vs 17.2±0.7 kg, respectively, P=0.07). Pasture allowance did not affect any recorded behaviour (mean ± sem for HPA and LPA respectively: number of times that the lamb entered the area: 6.2±0.6 vs 5.6±0.9; time spent within the area: 214.2±12.9 s vs 203.5±17.3 s; vocalizations: 25.0±2.4 vs 28.7±3.8; attempts to reunite with their mother: 15.5±1.6 vs 10.0 vs 2.5). Low natural pasture allowance from the beginning of gestation until lambing did not affect the response of the lambs to the short-term separation from their mothers when they were 3 months old.

Shearing pregnant ewes in winter in a Mediterranean environment increases vigour of the offspring

David Miller, Mara Farrelly and Caroline Jacobson
Murdoch University, School of Veterinary and Life Sciences, 90 South St, Murdoch WA 6150,
Australia; d.miller@murdoch.edu.au

Shearing ewes in mid-late pregnancy in temperate environments has been shown to increase lamb survival rates under both pastoral conditions and winter-housed environments. However, it is unknown if this effect can occur under the relatively mild winter conditions of a Mediterranean climate. The hypothesis tested was that shearing ewes in mid-late pregnancy increases the vigour of lambs born in the Mediterranean environment of south Western Australia. Twenty pregnant Poll Dorset ewes were randomly stratified into two equal groups with similar live weights and body condition score. Ten ewes were shorn and ten were sham-shorn (unshorn) on day 100 of pregnancy in July (mid-winter). The sham-shorn sheep underwent the same handling process of shearing as the shorn sheep, but without removing the fleece. The mean minimum temperature for the 10 days immediately following shearing was 3.9 ± 1.3 °C ($\pm$ SEM) and the mean maximum temperature was 17.7 ± 0.5 °C. Analysis of blood samples collected from the ewes 24 hours after shearing showed that shearing resulted in raised plasma cortisol concentrations (72 ± 12 ng/ml vs 41 ± 8 ng/ml) in the shorn ewes compared to the unshorn ewes ($P<0.05$). All the pregnant sheep were handled frequently prior to lambing so the ewes were accustomed to human presence for observation. Three days prior to expected lambing the sheep were moved to a smaller observation area. Immediately after birth a blood sample was quickly collected from the lamb so as not to disturb maternal-newborn bonding, and then behavioural measurements related to lamb vigour were collected. Length of parturition and birth weight were not different between the two treatment groups. Lambs from shorn ewes took 25% less time to suckle after birth ($P<0.05$) and they also took 35% less time to stand after birth ($P<0.05$), compared to the lambs from unshorn ewes. From the blood sample taken from the lambs immediately after birth, analysis revealed no difference in circulating cortisol, glycerol, triglyceride or non-esterified fatty acid concentrations between the two treatment groups. The lambs from the unshorn ewes had 91% higher circulating glucose concentrations than the lambs from the shorn ewes ($P<0.05$), which may indicate greater reliance on mobilisation of glycogen stores rather than brown adipose tissue for thermogenesis in the lambs from the unshorn ewes. In conclusion, lambs displayed better vigour when the maternal ewe was shorn in mid-late pregnancy in the winter of a Mediterranean climate. It is possible that the changes in the vigour of the lambs from the shorn ewes, and the difference in circulating glucose concentration, may have arisen by fetal programming effects on thermogenic systems resulting from the cold stress (fleece removal) that affected the maternal glucocorticoid system.

Can we increase calves' milk intake during the first days of their lives by facilitating their normal sucking behaviour?

Anne Marie De Passille and Jeffrey Rushen
University of British Columbia, Main Mall, Vancouver, BC, Canada; amdepassille@gmail.com

Dairy calves can drink prodigious quantities of milk during their first few days of life, but there are large differences between calves, and a low milk intake is associated with increased risk of illness and poor growth later in life. We examined whether we could increase the milk intake of peri-parturient calves by a better positioning and design of the milk feeder to facilitate the calves' normal sucking behaviour. Holstein calves (1-5 d old) were housed individually and given 8 l of milk twice a day through a teat attached to a milk reservoir. In experiment 1, we placed the milk reservoir for 12 calves at two heights above the teat (50 and 135 cm) and measured milk intake, latency to drink milk, duration of assistance needed and weight gain. Milk intake over the 4 d was higher for calves fed from the elevated reservoir (Wilcoxon tests: $P=0.02$) but the effect varied from meal to meal: the largest effects were noted on the 3rd meal (medians = 2.9 vs 1.9 l; $P=0.04$) and the 5th meal (4.9 vs 2.4 l; $P=0.02$). There were no effects on weight gain or the amount of assistance given. In experiment 2, the teat was held either in a plain wooden board or in a board covered with a soft material to facilitate the calves' (n=28) butting behaviour. On each day, the calves with the covered board drank more milk but the effect was not significant ($P>0.10$). However, the variation between calves in milk consumption was reduced with no calves with the covered board drinking less than 4 l per meal. There were no significant effects on overall weight gain ($P>0.10$) but there were inconsistent effects at each day: calves with the covered board gained more weight on days 2 and 4, but less weight on d3 ($P<0.05$). There was a trend ($P=0.07$) for a reduced latency to first drink after milk delivery for calves with the covered board (median = 40 min vs 81 min). Increasing the height of the milk reservoir above the teat reduces the amount of effort needed for the calf to suck the milk, and increases milk intake. Covering the teat holder with a soft material to facilitate butting does not appear to increase overall milk intake or weight gains but may be helpful for smaller or weaker calves. More attention should be paid to the design and positioning of milk feeding equipment to improve early milk intakes by dairy calves.

Short and long term effects of free and half-day contact between dairy cows and their calves

Katharina A. Zipp, Yannick Rzehak and Ute Knierim
University of Kassel, Farm Animal Behaviour and Husbandry Section, Nordbahnhofstr. 1a, 37213 Witzenhausen, Germany; zipp@uni-kassel.de

In dam rearing systems dairy cows nurse their calves and are additionally milked. The consequent loss of sellable milk is likely affected by the daily duration of cow-calf-contact. The influence of three different calf rearing systems on machine milk yield and content, calf and heifer development were investigated in German Black Pied dairy cows: 'free' (24 h, n=13), 'half-day' (6:45-18:00 h, n=11) and no contact ('control', n=14) with the dam during the first nine weeks of life. 'Control' calves were separated from the dam after birth and fed max. 2×3 l/d whole milk with nipple buckets. All cows were milked twice daily. During the 10th week of life, dam reared calves were separated from the cows, trained to drink from nipple buckets and fed as 'control' calves. All calves were gradually weaned until the 13th week of life. Due to missing weighings and as male calves were sold after the 10th week sample size varied. Machine gained milk yield of dams during the suckling period was significantly lower than in 'controls' ('free': 8.5±2.9, 'half-day': 12.2±4.6 vs 22.1±4.4 kg/d, mixed effects model, P<0.05). 'Half-day' cows had in tendency (P=0.0758) higher daily milk yields than 'free' cows. This was also true for the whole lactation (ANOVA, 'free': 14.1±2.6 kg/d n=11, 'half-day': 16.7±3.7 kg/d, P=0.0889, n=10), and 'half day' did not significantly differ from 'control' any longer (18.5±2.1 kg/d, P=0.2193 n=12). Lower fat contents in dams during the suckling period were a sign of disturbed milk ejection ('free': 3.34±0.21%, 'half-day': 3.10±0.38 vs 3.96±0.24%, mixed effect model, P<0.005). No differences in Somatic Cell Score could be detected (P>0.1). Daily weight gain did not differ among dam reared calves, but was significantly higher than in 'control' calves during the suckling period (ANOVA, 'free': 0.97±0.14 kg/d, n=10, 'half-day'=0.97±0.14 kg/d, n=7, 'control': 0.64±0.08 kg/d, n=12, P<0.00 1). The week after separation from the dam there was a growth check in both dam reared groups ('free': 0.43±0.20 kg/d, n=12, 'half-day'=0.36±0.45 kg/d, n=9, 'control': 0.83±0.27 kg/d, n=11, P<0.05), but two weeks after weaning body weight was still higher than in 'controls' (Kruskal-Wallis and Wilcoxon rank sum test, median±median absolute deviation, 'free': 120.5±11.8 kg, n=8, 'half-day'=128.0±9.5 kg, n=5, 'control': 109.3±6.8 kg, n=8, P<0.05). Later, when integrated into the milking herd as heifers after calving (n_{free}=5, $n_{half-day}$=5, $n_{control}$=9), body weight, height at withers, trunk girth, age at calving and milk production did not differ between groups. During the first 24 h, all heifers showed considerably reduced lying durations (recorded with Onset Pendant G data loggers; Onset Computer Corporation, Bourne, MA). During the second 24 h 'free' lay significantly and 'half-day' in tendency longer than 'controls' (Kruskal-Wallis and Wilcoxon rank sum test, median±median absolute deviation, 'free': 6.8±0.3 h, 'half-day': 6.2±0.7 h, 'control': 4.6±0.9 h) due to more lying bouts. Results suggest that half-day contact may be a way to reduce the loss of sellable milk without losing advantages of a dam rearing system. Ways to prevent a growth check after separation from the dam are needed.

Fence-line separation of dairy calves at four days and its effect on behaviour

Lena Lidfors[1], Anya Törnkvist[1], Emma Fuxin[1] and Julie Johnsen Foske[2]
[1]Swedish University of Agricultural Sciences, Department of Animal Environment and Health, P.O. Box 234, 532 23, Sweden, [2]National Veterinary Institute, Postboks 750 Sentrum, 0106 Oslo, Norway; lena.lidfors@slu.se

The purpose of this study was to investigate if cows and calves would show less behavioural stress responses when separation after the colostrum period was done so that they could see and touch each other (Fence-Line FL) instead of just hearing each other (Hearing Contact HC). The study was done on an organic farm with about 85 dairy cows of the Swedish Red and Swedish Holstein breed. The cows were kept in a cubicle system with a Voluntary Milking system. Cows were moved into individual calving pens adjacent to the cubicle housing just before calving and all calves were given 2 l of colostrum in a bottle within 6 h after birth. Cow-calf pairs (n=8/ treatment) were separated at 3-4 days after birth when cows were released into the cubicle system and their calves were placed in one of the treatment pens. FL separation had a pen placed in the middle of the cubicle system with two walls and two gates where the cows and calves could touch and see each other but the calf could not suckle the cow. HC separation had the calves moved to an adjacent pen which had a plastic curtain so that the mother could not see or touch her calf, but she could hear it. Calves were fed 3 l of milk in teat buckets just prior to observations. Cow and calf were recorded during 0-2 and 24-28 h after separation with instantaneous recordings at 5 min intervals of lying and continuously of other behaviours. Behaviours are shown as mean ± SE frequency/h or % of observations. Data were analyzed with a Generalized Linear Model for the effect of treatment, time of observation and the interaction. Calves with FL separation performed a higher frequency of play behaviours (5.5±1.71) than calves with HC (0.2±0.11, P<0.001), and play tended to be performed more often 24-28 h than 0-2 h after separation (P=0.08). Calves with FL separation kept their head through the fence a higher frequency (3.8±0.94) than calves with HC separation (1.3±0.58, P<0.05), but there were no differences between days (P=0.90). For high pitch vocalization in calves there was a tendency of a difference between observation times (P=0.08), but no effect of treatment (P=0.14; FL 0-2 h: 1.7±1.44, FL 24-28 h: 5.8±3.22, HC 0-2 h: 6.0±4.24, HC 24-28 h: 15.0±11.83). High pitch vocalization in cows tended to differ between observation times (P=0.097), but not between treatments (P=0.16; FL 0-2 h: 9.1±3.09, FL 24-28 h: 3.9±2.36, HC 0-2 h: 15.3±6.93, HC 24-28 h: 11.3±6.28). Low pitch vocalization did not differ between treatments in either cows (P=0.76, 4.2±0.98) or calves (P=0.22, 1.6±0.73), nor between observation times in cows (P=0.17) or calves (P=0.17). Cows were lying down a higher percentage of the observations at 24-28 h (19.5±5.84) than at 0-2 h after separation (6.0±3.43, P<0.0001), but there were no differences between FL and HC (P=0.75). Calves were lying down 60.3±5.10% of the observations and it did not differ between treatments (P=0.72) or observation times (P=0.38). In conclusion calves appeared to be more activated by being placed in physical contact with the cow at separation, but there were no clear effects on the cows stress level.

Effect of calf suckling, weaning and use of calf dummy as milk ejection stimuli on behaviour and performance of Sahiwal cows

M.L. Kamboj, A. Kumar, P.K. Singh, N. Kumar, M. Saini and S. Chandra
National Dairy Research Institute (NDRI), Livestock Production Management Section, Karnal (Haryana), 132 001 Karnal (Haryana), India; kamboj66@rediffmail.com

Traditional smallholder farmers in India predominantly use suckling by the calf before milking as a milk ejection stimulus in indigenous breeds of cows and buffaloes. However, at many large farms, calves are weaned at birth and milk let-down stimulus is provided by offering concentrate mixture and teat massage before milking. This is presumed to affect milking temperament and result in poor milk ejection and shorter lactations. In suckled animals, death or separation of the calf causes a serious milk ejection problem and to avoid it some farmers resort to indiscriminate use of exogenous oxytocin. Using a synthetic dummy calf after bonding with dams soon after parturition was hypothesized to ward off adverse effects of both suckling and weaning. The aim of our study was to assess effects of 3 different milk ejection stimuli on milking behaviour and performance of cows and buffaloes. For this, 24 late-pregnant Murrah buffaloes and 18 Sahiwal cows (1-4 parity) were blocked by milk yield in previous lactation and randomly allocated to one of 3 treatments for 6 months from calving (in each treatment n=8 for buffaloes and n=6 for cows). Treatment 1: animals suckled by their calves before milking (S), Treatment 2: calves weaned at birth and animals offered concentrate mixture before milking (W), Treatment 3: animals presented with a dummy calf before milking (D). The animals were hand milked and milk yield was recorded daily using electronic weighing balance. Milking behaviour parameters were recorded manually at weekly intervals. Data were analysed using one way ANOVA in SPSS for testing significance of differences among treatment means. Mean milk let-down time in S, D and W cows respectively was 0.86 ± 0.02, 1.27 ± 0.05 and 1.74 ± 0.02 min and in buffaloes 1.87 ± 0.13, 2.40 ± 0.18 and 3.62 ± 0.26 min; being statistically similar for S and D and significantly ($P<0.01$) higher in W. Mean milking temperament score (1-5 scale, Tulloh, 1961) in buffaloes was similar for treatments W (2.69 ± 0.17) and D (2.14 ± 0.23) but significantly ($P<0.05$) lower (1.38 ± 0.13) in S. In cows these values were similar in S (1.29 ± 0.43) and D (1.57 ± 0.22) but significantly ($P<0.05$) higher in W (2.12 ± 0.19). Mean total lactation yields in cows ($2,568\pm232$, $2,108\pm329$ and $1,770\pm241$ kg) and in buffaloes ($2,785\pm279$, $2,211\pm240$ and $1,739\pm264$ kg in S, D and W respectively) were significantly ($P<0.01$) different among 3 treatments. Mean lactation lengths in S, D and W respectively in cows were 336 ± 18, 285 ± 19 and 270 ± 29 d and in buffaloes 349 ± 31, 297 ± 20 and 256 ± 19 d; being highly significantly different ($P<0.01$) in both species. We concluded that milking behaviour was almost comparable in suckled and dummy-calf-used Sahiwal cows and Murrah buffaloes and productive performance was the highest in suckled animals followed by dummy-calf-used animals. Weaning had the most adverse consequences on milking behaviour and performance of these animals.

The natural weaning window of suckler beef cattle

Dorit Albertsen and Suzanne Held
University of Bristol, School of Veterinary Sciences, Langford House, Langford, Bristol, BS40 5DU,
United Kingdom; dorit.albertsen@bristol.ac.uk

Leaving yearling calves with their mothers to be weaned naturally is the husbandry method on an extensive UK beef farm, aimed at avoiding the negative impacts of cow-calf separation on calf health. The focus of this study was to determine the natural weaning window for domestic beef cattle. It was suspected that cows would wean their calves after around 80% of the time available between parturitions. The udders of 55 Aberdeen Angus cows were inspected visually and scored for signs of suckling on a weekly basis between September and May over five rearing seasons. On the date a cow was first found un-suckled, the cow's calf was declared weaned if every subsequent visit confirmed that no further suckling had occurred. An average, and the range, of days from birth to weaning were taken to determine the weaning window in days, and because of the variation in calving interval also in percent of time available. The data set was investigated first analysing it per cow (n=55), averaging her values, and again per calf (n=133), including incomplete repeated measures for repeatedly monitored cows (n=38). Linear regressions were used to investigate the relationship between the calving interval and the weaning window. The effect of cow age on calf weaning age was tested with ANOVA. The influence of calf gender on weaning age was tested with Independent Samples T-tests. 128 (96.2%) calves were weaned by their mother. Weaning averaged at 9.7 months (294 days, SD±34.35 days) post-partum within a mean calving interval of 363 days, for the 53 cows who weaned. These cows spent 82.5% (SD±6.85%) of the time between parturitions nursing their calf at foot. Linear regression revealed a significance for cows with a longer calving interval to wean older calves ($R^2=0.171$, df=1, P<0.001), spend a smaller percentage of the calving interval nursing their calf ($R^2=0.048$, df=1, P=0.01) and spend less time suckling while pregnant, thereby reducing the amount of time they were investing simultaneously into the foetus and the calf at foot ($R^2=0.114$, df=1, P<0.001). For the year 2012/13 One-Way ANOVA showed significant differences in the time to weaning between older and younger cows (F=4.02, df=2, P=0.032), in 2013/14 it did not (F=1.708, df=2, P=0.2). In four out of the five observation seasons the average weaning age for female calves was lower than that for male calves, although the difference was only significant in 2009/10 ($t_{(13)}$=-3.1, P=0.009). The majority of suckler beef cows observed in this study weaned their calves off milk before their next parturition. The natural weaning window shown by these cows has a flexible time range that appears to depend on the time available between parturitions, and might be influenced by cow age and calf gender. Behavioural studies are under way.

Authors index

Printed in the United States
by Baker & Taylor Publisher Services